Agent_Zero

Princeton Studies in Complexity

SERIES EDITORS

Simon A. Levin (Princeton University)

Steven H. Strogatz (Cornell University)

Lars-Erik Cederman, *Emergent Actors in World Politics: How States and Nations Develop and Dissolve*

Robert Axelrod, *The Complexity of Cooperation: Agent-Based Models of Competition and Collaboration*

Peter S. Albin, *Barriers and Bounds to Rationality: Essays on Economic Complexity and Dynamics in Interactive Systems.* Edited and with an introduction by Duncan K. Foley

Duncan J. Watts, *Small Worlds: The Dynamics of Networks between Order and Randomness*

Scott Camazine, Jean-Louis Deneubourg, Nigel R. Franks, James Sneyd, Guy Theraulaz, and Eric Bonabeau, *Self-Organization in Biological Systems*

Peter Turchin, *Historical Dynamics: Why States Rise and Fall*

Andreas Wagner, *Robustness and Evolvability in Living Systems*

Mark Newman, Albert-Laszlo Barabasi, and Duncan Watts, eds., *The Structure and Dynamics of Networks*

J. Stephen Lansing, *Perfect Order: Recognizing Complexity in Bali*

Joshua M. Epstein, *Generative Social Science: Studies in Agent-Based Computational Modeling*

John H. Miller and Scott E. Page, *Complex Adaptive Systems: An Introduction to Computational Models of Social Life*

Michael Laver and Ernest Sergenti, *Party Competition: An Agent-Based Model*

Joshua M. Epstein, *Agent_Zero: Toward Neurocognitive Foundations for Generative Social Science*

Agent_Zero

Toward Neurocognitive Foundations for Generative Social Science

Joshua M. Epstein

Princeton University Press

Princeton and Oxford

Published by Princeton University Press, 41 William Street, Princeton, New Jersey 08540
In the United Kingdom: Princeton University Press, 6 Oxford Street, Woodstock, Oxfordshire OX20 1TW

press.princeton.edu

Library of Congress Cataloging-in-Publication Data

Epstein, Joshua M., 1951–
Agent zero : toward neurocognitive foundations for generative social science / Joshua M. Epstein.
pages cm. — (Princeton studies in complexity)
Summary: "The Final Volume of the Groundbreaking Trilogy on Agent-Based Modeling In this pioneering synthesis, Joshua Epstein introduces a new theoretical entity: Agent Zero. This software individual, or "agent," is endowed with distinct emotional/affective, cognitive/deliberative, and social modules. Grounded in contemporary neuroscience, these internal components interact to generate observed, often far-from-rational, individual behavior. When multiple agents of this new type move and interact spatially, they collectively generate an astonishing range of dynamics spanning the fields of social conflict, psychology, public health, law, network science, and economics. Epstein weaves a computational tapestry with threads from Plato, Hume, Darwin, Pavlov, Smith, Tolstoy, Marx, James, and Dostoevsky, among others. This transformative synthesis of social philosophy, cognitive neuroscience, and agent-based modeling will fascinate scholars and students of every stripe. Epstein's computer programs are provided in the book or on its Princeton University Press website, along with movies of his "computational parables." Agent Zero is a signal departure in what it includes (e.g., a new synthesis of neurally grounded internal modules), what it eschews (e.g., standard behavioral imitation), the phenomena it generates (from genocide to financial panic), and the modeling arsenal it offers the scientific community. For generative social science, Agent Zero presents a groundbreaking vision and the tools to realize it"—Provided by publisher.
Includes bibliographical references and index.
ISBN 978-0-691-15888-4 (hardback)
1. Social sciences—Computer simulation. 2. Social sciences—Mathematical models. I. Title.
H61.3.E6697 2013
300.1—dc23 2013018009

British Library Cataloging-in-Publication Data is available

The research for this book was funded by a 2008 NIH Director's Pioneer Award

This book has been composed in Minion Pro and Fairfield

Printed on acid-free paper. ∞

Printed in the United States of America

1 3 5 7 9 10 8 6 4 2

For Matilda,
My lark at break of day arising . . .

Contents

Foreword xi
Preface xiii
Acknowledgments xv

Introduction 1
MOTIVATION 1
Generate Social Dynamics 2
A Core Target 2
THE MODEL COMPONENTS 5
MODEL OVERVIEW 6
Skeletal Equation 8
Specific Components 9
ORGANIZATION 10
Part I: Mathematical Model 10
Part II: Agent-Based Model 11
Part III: Extensions 13
Replicability and Research Resources on the Princeton University Press Website 16
Part IV: Future Research and Conclusions 17

Part I. Mathematical Model 19
I.1. THE PASSIONS: FEAR CONDITIONING 19
Fear Circuitry and the Perils of Fitness 20
Nomenclature of Conditioning 29
The Rescorla-Wagner Model 33
Social Examples 37
Fear Extinction 41
I.2. REASON: THE COGNITIVE COMPONENT 46
I.3. THE SOCIAL COMPONENT 51
Simple Version of the Core Target 55
Examples of Fear Contagion 57
Mechanisms of Fear Contagion 59

Conformist Empirical Estimates 63
Generalizing Rescorla-Wagner 67
The Central Case 69
Tolstoy: The First Agent Modeler 71
A Mathematical Aside on Social Norms as Vector Fields 74
Extinction of Majorities 78
I.4. INTERIM CONCLUSIONS 80

Part II. Agent-Based Computational Model 81
Affective Component 84
"Rational" Component 85
Social Component 88
Action 89
Pseudocode 89
II.1. COMPUTATIONAL PARABLES 90
Parable 1: The Slaughter of Innocents through Dispositional Contagion 90
Parable 2: Agent_Zero *Initiates: Leadership as Susceptibility to Dispositional Contagion* 94
Run 3. Information Cuts Both Ways 96
Run 4. A Day in the Life of Agent_Zero*: How Affect and Probability Can Change on Different Time Scales* 98
Run 5. Lesion Studies 102

Part III. Extensions 107
III.1. ENDOGENOUS DESTRUCTIVE RADIUS 107
III.2. AGE AND IMPULSE CONTROL 109
III.3. FIGHT VS. FLIGHT 110
Case 1: Fight 111
Case 2: Flight 112
Capital Flight 114
III.4. REPLICATING THE LATANÉ-DARLEY EXPERIMENT 114
Threshold Imputation 115
The Dialogue 118
III.5. MEMORY 118
III.6. COUPLINGS: ENTANGLEMENT OF PASSION AND REASON 122
Mathematical Treatment 124
III.7. ENDOGENOUS DYNAMICS OF CONNECTION STRENGTH 128
Affective Homophily 128

General Setup 130
Agent-Based Model: Nonequlibrium Dynamics 135
III.8. GROWING THE 2011 ARAB SPRING 138
III.9. JURY PROCESSES 143
Phase 1. Public Phase 143
Phase 2. Courtroom Trial Phase 145
Phase 3. Jury Phase 147
III.10. EMERGENT DYNAMICS OF NETWORK STRUCTURE 152
Network Structure Dynamics as a Poincaré Map 153
Relation to Literature 159
III.11. MULTIPLE SOCIAL LEVELS 160
Agent_Zero *as Witness to History* 161
III.12. *THE 18TH BRUMAIRE OF* AGENT_ZERO 165
III.13. INTRODUCTION OF PRICES AND SEASONAL ECONOMIC CYCLES 168
Prices 168
A Christmas Story 173
III.14. SPIRALS OF MUTUAL ESCALATION 176

Part IV. Future Research and Conclusion 181
IV.1. FUTURE RESEARCH 181
IV.2. CONCLUSION 187
Civil Violence 187
Economics 188
Health Behavior 189
Psychology 190
Jury Dynamics 191
The Formation and Dynamics of Networks 191
Mutual Escalation Dynamics 192
Birth and Intergenerational Transmission 192
IV.3. TOWARD NEW GENERATIVE FOUNDATIONS 192

Appendix I. Threshold Imputation Bounds 195
Appendix II. Mathematica *Code* 197
Appendix III. Agent_Zero NetLogo *Source Code* 213
Appendix IV. Parameter Settings for Model Runs 221
References 227
Index 243

Foreword

I SEE THIS BOOK as the third in a trilogy on generative social science.

VOLUME I

The first volume of the trilogy was *Growing Artificial Societies: Social Science from the Bottom Up* (MIT Press/Brookings Press), with coauthor Robert Axtell. Published in 1996, this introduced the *Sugarscape* agent-based model, and the notion of a *generative explanation* of social phenomena. Sugarscape was a single sweeping exploratory artificial society, with glimmerings of a mature generative epistemology.

VOLUME II

For the subsequent decade, with diverse colleagues, I applied agent-based modeling to a broad spectrum of fields—economics, archaeology, conflict, epidemiology, spatial games, and the dynamics of norms—and thought more deeply about the philosophy of agent-based social science. The results are collected in the second volume of the trilogy: *Generative Social Science: Studies in Agent-Based Computational Modeling,* published in 2006 by Princeton University Press. Relative to *Growing Artificial Societies, Generative Social Science* presented a collection of more tightly focused and specifically explanatory exercises, and a far more extended and mature generative epistemology.

VOLUME III

Generative Social Science ended with a challenge: *Grow Raskolnikov*! By this, I meant agents with more fully developed—and so conflicted—inner lives.[1] In effect, it was a call for greater cognitive realism. The present book is my

[1] Indeed, the name Raskolnikov is from the Russian, *raskól'nik,* meaning, roughly, schismatic.

response. It introduces a new theoretical entity, *Agent_Zero*, whose observable behavior is itself generated by the interaction, indeed conflict, of affective, deliberative, and social components. Passion, reason, and social forces, in other words, are all at play in *Agent_Zero*'s observable behaviors, which span a wide array of fields, including economics, health, conflict, social psychology, and endogenous network dynamics. Needless to say, the relationship between passion and reason within the individual, and the relation of the individual to society, are among the more enduring questions in philosophy. And while I do not claim to resolve them, I do claim to treat them in a new way. A significant volume of contemporary cognitive neuroscience is employed in constructing *Agent_Zero*, whom I offer as a new, neurocognitively grounded, foundation for generative social science—hence the subtitle of this book. However, as is repeated numerous times, I use only selected neuroscience in developing this particular agent model. I do not purport to encompass—much less to advance—any area of neuroscience itself. Indeed, through the good offices of a number of very fine neuroscientists, I may have avoided insulting their fascinating and fast-moving discipline.

Preface

EMOTIONAL, COGNITIVE, AND SOCIAL FACTORS shape the behavior of individuals in groups and hence shape the emergence of important social dynamics, from genocide to financial panic. I wish to generate such social dynamics "from the bottom up," in social networks of neurocognitively plausible individuals. To this end, I introduce a new theoretical entity, *Agent_Zero*, endowed with interacting emotional/affective, cognitive/deliberative, and social modules. *Agent_Zero*'s affective component is based on the Rescorla-Wagner model of conditioning and extinction, supported by recent science on the neural mechanisms of fear conditioning specifically. The agent's cognitive (deliberative) component reflects well-documented biases and heuristics in the estimation of probabilities (e.g., sample selection bias). Agents belong to social networks, and the social component exhibits contagion effects. But, crucially, it is not observable behavior that is transmitted in this model, but *disposition*. I define this here as an explicit function of (a) the individual's emotion and cognition ("passion and reason"), and (b) others' affective and deliberative states. Action is binary and is triggered when individual disposition exceeds threshold. These thresholds, and susceptibility to dispositional contagion, can be heterogeneous across agents, all of whom can exhibit emotional inertia and memory of events. The same basic model, interpreted and extended variously, is shown to generate core phenomena in the fields of social conflict, psychology, public health, law, network dynamics, and economics. Mathematical and spatial agent-based computational versions of the general model are presented.

I believe *Agent_Zero* to be a departure in what it includes (e.g., neurally grounded internal modules), what it excludes (e.g., standard behavioral imitation), the range of phenomena it generates, and the set of tools it offers the field. Overall, I submit *Agent_Zero* as a step toward unified—and neurocognitively grounded—foundations for generative social science.

Acknowledgments

THIS RESEARCH WAS SUPPORTED entirely by a National Institutes of Health Director's Pioneer Award (DP1). This afforded me unparalleled scientific freedom and resources to found a new Center for Advanced Modeling in the Social, Behavioral, and Health Sciences at Johns Hopkins University, where this research was primarily conducted. For these exceptional opportunities, I am eternally grateful to the NIH. The Center would not have been possible without the vision and support of Dr. Gabor Kelen, Chairman of the Department of Emergency Medicine, to whom I am profoundly grateful. I also wish to thank my colleagues in the department for the climate of uncompromising excellence in which it is my privilege to work. My closest colleague in this project has been Julia Chelen, whom I thank for superb research assistance and invaluable consultations throughout. For detailed reviews of the manuscript, I am indebted to Robert Axelrod, Jon Parker, and Lisa Feldman-Barrett. For important reactions, from a variety of scientific perspectives, I also thank Paul Smaldino, Eili Klein, Michael Makowsky, David Broniatowski, and Erez Hatna. I am grateful for numerous sessions on learning theory with Peter Holland. For many formative discussions and for cohosting a two-day NIH Workshop on the model, I thank John Cacioppo. For their many insights and comments, I am grateful to the workshop participants: Colin Camerer, Julia Chelen, William Eaton, Stephen Eubank, Thomas Glass, Joseph Harrington, Melissa Healy, Peter Holland, Gabor Kelen, Ethan Kross, Jon Parker, Elizabeth Phelps, Paul Torrens, and Richard Rothman. I am deeply indebted to the Princeton University Press external reviewers Duncan Foley, Paul Slovic, and John Steinbruner for thorough readings, penetrating insights, and valuable suggestions for improvement. I also thank the book's Princeton University Press editorial and production team for their engagement and expertise: Vickie Kearn, Mark Bellis, Dimitri Karetnikov, and Linda Thompson. I thank Donald Burke for many stimulating colloquies about the possibility of a unified behavioral health science. The first year of this research was conducted at The Brookings Institution, which is gratefully acknowledged. Ross Hammond, Kislaya Prasad, and Peyton Young offered valuable comments during this phase of the research. For countless discussions of generative minimalism and the philosophy of science, I thank my brother, Samuel David Epstein. Finally, for their love and support, I thank my wife, Melissa Healy, our son, Joey, and our daughter, Matilda, to whom this book is dedicated, in memory of my parents, Joseph and Lucille Epstein.

Agent_Zero

INTRODUCTION

MOTIVATION

IN HIS *TREATISE OF HUMAN NATURE*, David Hume (1739; 2000 ed.) famously wrote, "Reason is . . . the slave of the passions."[2] In using the term *slave*, however, Hume's point is not that the passions always prevail over reason in a tug of war,[3] but that the two are, in some sense, incommensurable. Distinguishing the passions from factual (true/false) claims, he writes, "Tis impossible [that] they [the passions] can be pronounced either true or false, and be either contrary or conformable to reason" (p. 458). I take this "nonconformability" to mean that, as a modeling proposition, passion and reason—emotion and ratiocination, if you prefer—belong on different axes. They might be seen as basis elements of a dispositional vector space. If we adopt Hume's geometry, then beyond passion and reason there is a third basis element, or axis, inspired by Spinoza, who wrote that "Man is a social animal."[4] In this view, then, our actions may be motivated by the passions, but are influenced by reason—and by society.

Needless to say, Hume was not the last, nor Spinoza the first, to philosophize on the roles of emotion, deliberation, and social influence in human behavior. Indeed, innumerable contrasting perspectives on the balance among them have been articulated over the centuries. I do not purport to end this dialogue. I do claim, however, to inject into that discussion a new approach.

Specifically, my central objective here is to develop a simple explicit model of individual *behavior in groups* that includes some representation of "the passions," of (imperfect) reason, and of social influence. In other words, I will offer an exploratory synthesis of three (partially understood and obviously intertwined) processes:

[2]The entire passage is, "Reason is, and ought only to be, the slave of the passions." His use of the term *ought* is, of course, not an admonition, but carries the meaning that it is in the nature of things that reason be the slave of the passions, a point that struck me as incidental to the main topic at hand. Hence the ellipses.

[3]On the contrary, he writes, "We speak not strictly and philosophically when we talk of the combat of passion and reason." That is, we speak incorrectly when we do (Vol. 2, p. ix).

[4]Baruch Spinoza (1632–1677). *Ethics*, Pt. IV, prop. 35: note. The addage is often attributed to Aristotle. But, technically, he wrote that, "Man is by nature a political animal." *Politics 1.* One can argue that "social" is a defensible translation of the original Greek, *politikos,* in that the latter means "of the polis" and for Aristotle, the polis *was* society. But with Spinoza there is no doubt.

- The emotional
- The cognitive
- The social

To this end, I introduce a new theoretical entity, *Agent_Zero*, endowed with emotional/affective, cognitive/deliberative, and social modules whose—often nonconscious—interactions determine his or her observed behavior.

Generate Social Dynamics

The *aim* is to *generate* recognizable dynamics of social importance. Specifically, we humans often do things in groups that we would not do alone (Le Bon, 1895; Canetti, 1984; Mackay, 1841, Browning, 1998). We do things for which we have no basis in evidence. Indeed, we do them *knowing* we have no evidence; and sometimes, despite this, we are even the *first* in the group to do them!

Dispositional Contagion, Not Behavioral Imitation

Notice that in this latter case, the *imitation of behavior* cannot be the mechanism. Clearly, if I am the first actor, I cannot be imitating anyone's behavior, because no one before me has acted! But, more to the point, in this model my successors are not imitating my behavior either. No agent is imitating any other's *behavior* as defined in this model. Here, behavior is binary—agents act or do not—and, by design, this Boolean variable does not appear as an independent (right-hand side) variable in any agent's algorithm (discussed more fully later). Of course, agents may end up *doing* identical things, but the mechanism introduced here is *dispositional contagion*, not "monkey see, monkey do" imitation of observable binary action, which, as I say, is barred mathematically. Without positing "behavioral imitation" in the usual sense,[5] the *Agent_Zero* model will be shown to generate a wide range of social dynamics.

A Core Target

As a core target for the model, I want to "grow" the person who *feels* no aversion to black people, who has never had any direct *evidence* or experience of black wrongdoing (his empirical estimate of this probability is and always has been literally zero), and who yet initiates the lynching. Though,

[5] Behavioral imitation can be a productive modeling assumption. I have used it myself. See, for example, Chapters 7 and 10 of J. M. Epstein (2006). It is not used here.

as first mover, he does not copy the *behavior* of others, a type of contagion is at play, but it is dispositional. He is, of course, an ideal type, but an important one.[6]

Mathematical social science has largely attempted to characterize the rational; I am interested in generating behavior that is far from rational. And I believe there is more at play than mere bounds on information or computing resources. While cases of collective cruelty abound, there are also examples of collective resistance to it. A general theory should generate both. I will attempt to do so here.

Specifically, *Agent_Zero*'s observable behavior will result from the interplay of (a) emotional/affective, (b) cognitive/deliberative, and (c) network/contagion components. As defined here, the agent's total *disposition* to act will be an explicit mathematical function of these. This disposition will be a real number. If it exceeds some threshold, *Agent_Zero* acts. Otherwise she does not.

Each model component (affective, deliberative), of course, is an entire discipline in its own right. And the crude models set forth will doubtless merit criticism from domain experts. But I am less interested in the accuracy of the components than in the generative capacity of the synthesis. The components can—indeed must—be refined. But, as far as I know, this particular synthesis has not been attempted. So, to me, *it is less important to get the components finished than to get the synthesis started*, as simply and understandably as possible.

While it impinges on a number of enduring philosophical issues, this synthesis aims to fill an important gap in contemporary social science. In statistical social science, relationships among aggregate macroscopic variables are assessed econometrically. The individual, as such, is not represented. In mathematical economics and game theory, the individual *is* represented, but the representation is not grounded in cognitive neuroscience.[7] But laboratory neuroscience—transformed by advances in imaging—is focused on single individuals (e.g., fMRI subjects in controlled laboratory

[6] I do not study lynching empirically. For "black" and "lynching," one should feel free to substitute "Jew" and "pogrom," "Congolese civilian" and "slaughter," or any number of other bleak equivalents. Indeed, the endless supply of bleak equivalents is among the main things I wish to understand. The most recent confirmed lynching in the United States occurred in 1981, when ". . . an African-American teenager named Michael Donald was murdered by two members of the Ku Klux Klan, who slit his throat and hung his body from a tree in Mobile, Alabama. "*The 'Last Lynching': How Far Have We Come?*" Ted Koppel. Transcript available at http://www.npr.org/templates/story/story.php?storyId=95672737, National Public Radio (2008).

[7] The exciting new field of neuroeconomics (Glimcher et al., 2008) is commendably attentive to brain science. But it has been focused largely on individual economic decisions, not broader social behaviors, like violence, discrimination, or contagious fear, which are explored here.

experiments), not on *networks of* individuals.[8] So, what happens in *networks of cognitively plausible individuals? Indeed, by what mechanisms do cognitively plausible individuals generate networks?* This book explores these questions mathematically, and computationally—in the latter case, by the method of agent-based modeling.

Generative Minimalism

This effort is exploratory and unapologetically theoretical. It is an exercise in generative minimalism. I do not "fit" the basic model or its extensions to numerical data, though this is an obvious candidate for future research, nor do I claim to predict events.[9] Rather, I hope to demonstrate that the *Agent_Zero* model is *sufficient to generate* core qualitative social dynamics across a wide range of domains. These demonstrations establish the model's generative explanatory candidacy, in the language of J. M. Epstein (2006). For a full discussion of the generative explanatory epistemology, generative minimalism, and examples of empirically calibrated agent models, see J. M. Epstein (2006).[10]

Not Modeling Brain Regions

There is another, extremely important, qualification. We will, of course, have occasion to discuss a number of brain regions: the amygdala, hippocampus, and prefrontal cortex, for example. But I am emphatically *not modeling brain regions*. To clinicians, brain regions are of interest because functional localization facilitates therapy (e.g., neurosurgery). But to social theorists brain tissue is interesting exactly and only in so far as it licenses model design or model interpretation.

What is plausible for humans and what is not? The neuroscience will both illuminate and explain those limits, limits that belong in the model and that have been largely ignored by social science. The brain discussions that follow are meant only to support that modeling aim. If my account of the tissue science is wanting (as it surely is) or if the state of tissue science advances (as it surely will), I do *not* believe that per se endangers the mathematical model. One wants to develop a mathematical/computational framework that can exploit and accommodate the evolving

[8] Emerging efforts in hyperscanning take fMRIs of two interacting people simultaneously (Lee, Dai, and Jones, 2012; Montague et al., 2002).

[9] For a large variety of possible modeling goals other than prediction, see J. M. Epstein (2008).

[10] On generative minimalism, see also S. D. Epstein (2000), S. D. Epstein and T. D. Seely (2002, pp. 1–10) and S. D. Epstein and N. Hornstein (1999, pp. ix–xviii).

tissue science but is not hostage to today's snapshot of it, for surely there are few fields developing faster. At the same time, one's model should also be sufficiently binding as to *admit* falsification. I hope to strike this difficult balance.[11]

Thus, I will display the formal models I have in mind and try to undergird them with what little cognitive neuroscience science I know, hoping that the attempt will elicit the empathetic interest and collaboration of domain experts. Now to the model's formal constituents.

THE MODEL COMPONENTS

For the *affective/emotional* component, we will use the essentially Pavlovian theory of associative learning; specifically, a generalization of the Rescorla-Wagner (1972) model of conditioning. For the *cognitive*—ratiocinative and deliberate—component, agents will form a probability based on local sampling (which can introduce bias). For the *social* component, I will use a network transmission model, but with a crucial wrinkle. Most work on contagion in networks (including some of my own) posits the conscious observation and imitation of *behavior*. Very importantly, however, in the present model it is *not* observable (binary) behavior that is contagious in these networks, but dispositions. Finally, as a starting point, the function combining these affective, deliberative, and social components—the agent's executive function, if you like—will be the simplest imaginable: addition.[12] While numerous refinements are possible, the first issue is the generative capacity of this simplest of all possible syntheses.[13] As Einstein admonished, in modeling, *"everything must be made as simple as possible, but not one bit simpler"* (A Letter from the Publisher, 1962).

[11]Relatedly, it is clear that freezing in primal fear of the wriggling snake is different from cool taxonomic appraisal of the danger it might represent—different in the brain systems engaged and the physiological responses evoked. In a certain literature, these systems are sometimes nicknamed the "hot" and "cold" spheres of cognition. This terminology is used, for example, by the evolutionary psychologists Tooby and Cosmides (2008). Others use the terms *automatic* and *controlled* (Schneider, 2003). Stanovich and West (2000) introduced *System 1* and *System 2*. This terminology is adroitly employed by Kahneman (2011), who emphasizes that System 2 is "effortful," unlike System 1. I do not adopt any of these particular terminologies, which is not to criticize their use by the authors mentioned. I simply have a different objective here.

[12]The neural network literature offers a wealth of transfer functions *of* this sum. Classical backpropagation uses a sigmoid of this sum, for example. We will return to this, but for this initial development, we will simply *superpose* the components and compare to a threshold.

[13]Agents are endowed with memory only in the spatial agent-based model of Part II.

MODEL OVERVIEW

Though full details will be presented shortly, the basic mathematical scaffolding of the model is as follows. First, we imagine a binary action, $A \in \{0, 1\}$. Raid the icebox or don't. Participate in (or even initiate) the lynching or don't. Buy the BMW or don't. Accept a vaccination or don't. Support internment of Japanese Americans or don't. Flee the snake or don't. Wipe out the village or don't. *This is what we mean by "behavior." It is zero or one, a binary matter.* Exactly when do agents act in the model? This requires some steps.

Dispositions and the Action Rule

Agents will be endowed with specific affective and deliberative functions below. For the *i*th agent, these will be denoted $V_i(t)$ and $P_i(t)$. They will be dynamic and will change with experience. But at any time they return nonnegative real numbers between zero and one, inclusive, as values. We define the ith agent's *solo disposition* as simply the sum of these affective (V) and deliberative (P) components. So for Agent *i*.

$$D_i^{\text{solo}}(t) = V_i(t) + P_i(t). \qquad [1]$$

Agents also carry (unconsciously,[14] I assume) a set of numbers—*weights*—from the interval [0, 1] registering the influence of other agents' solo dispositions.[15] So, Agent 1 might be strongly influenced by Agent 2 but oblivious to Agent 3. If we let ω_{ji} denote the weight *of Agent j on Agent i*, then we would give ω_{21} a high value of perhaps 0.9 and ω_{31} a value of 0.

We next define the *total disposition* of Agent *i*, D_i^{tot}, as her own solo disposition plus the sum of the weighted solo dispositions of all other agents. Hence, at any time *t*,

$$D_i^{\text{tot}}(t) = D_i^{\text{solo}}(t) + \sum_{j \neq i} \omega_{ji} D_j^{\text{solo}}(t). \qquad [2]$$

This could, of course, be written as a single global summation with the convention that self-weights (ω_{ii}) are unity.[16]

[14] "Unconsciously" does not mean "while unconscious, as after a concussion," but simply "without awareness." I sometimes use the term nonconscious to avoid confusion.

[15] A single weight is applied to the full solo dispositions of others, not to their separate *V*s and *P*s. Obviously, the distributive law applies, but conceptually, I do not argue that agents have access to others' *V*s and *P*s, and of course the same solo disposition could be formed by an infinitude of *V*, *P* combinations.

[16] This will be our assumption. But agents might discount their own dispositions, setting $\omega_{ii} < 1$. They might also assign negative weights to the solo dispositions of other agents, a possible extension we will not explore here.

Now, each agent carries an *action threshold*: $\tau \geq 0$.[17] If one's total disposition D^{tot} exceeds the threshold, τ, then the action is taken: $A = 1$. Otherwise, it is not, and $A = 0$. In other words, agents act if and only if (denoted *iff*) their total disposition exceeds their threshold:

$$\text{Action Rule: Act iff } D_i^{tot}(t) > \tau_i. \quad [3]$$

If we further define the *i*th agent's *net disposition* $D_i^{net}(t)$ as $D_i^{tot}(t) - \tau_i$, the total net of threshold, then the agent's action rule can be stated succinctly as follows:

$$\text{Action Rule: Act iff } D_i^{net}(t) > 0. \quad [4]$$

In terms of our binary action variable, A, this is equivalently

$$\text{Action Rule: } A_i = \begin{Bmatrix} 1 \text{ if } D_i^{net} > 0 \\ 0 \text{ otherwise} \end{Bmatrix}. \quad [5]$$

In vector terms, if we employ a dot product[18] and the Heaviside unit step function,[19] H, Agent *i*'s binary action rule (equation [5]) can be written compactly as an equality:

$$\text{Action Rule: } A_i = H[\boldsymbol{\omega}_{ji} \cdot (\mathbf{V}_j + \mathbf{P}_j) - \tau_i]. \quad [6]$$

Here (suppressing time for clarity) $\mathbf{V}_j$ and $\mathbf{P}_j$, respectively, denote the vectors of all V and P values and $\boldsymbol{\omega}_{ji}$ denotes the vector of all weights *on* Agent *i*.[20]

Various formulations of total disposition, described shortly, will prove useful. But one of them is particularly revealing of the core distinction between behavioral imitation and dispositional contagion. We will refer to it as the *skeletal equation*, because it doesn't specify particular V and P functions.

[17] While the model allows heterogeneity, we will for the most part assume a single common threshold here.

[18] One type of vector is simply an n-tuple. Given two vectors, $\mathbf{A} = (a, b, c)$ and $\mathbf{B} = (x, y, z)$, their dot product $\mathbf{A} \cdot \mathbf{B}$ is $ax + by + cd$, the sum of position-wise products. On vector operations in general, see for example, Mardsen and Tromba, *Vector Calculus* (2011).

[19] The unit step function employed here is defined as follows: $H(x - y) = 0$ if $x \leq y$ and 1 if $x > y$. $H(x)$ equals 1 *if x strictly exceeds y* and equals zero otherwise. So, in [6], A_i equals 1 when the disposition dot product exceeds threshold. The Heaviside function is commonly used as a switching function in engineering, and is named after Oliver Heaviside, British mathematician. Some forms assign a third value, such as ½, to the case where $x = y$, and some variants use $H(x - y) = 0$ if $x < y$ and 1 if $x \geq y$. We use the specific definition given.

[20] So, for n agents, $\boldsymbol{\omega}_{ji} = (\omega_{1i}, \omega_{2i}, \ldots, \omega_{ni})$.

Skeletal Equation

Focusing on Agent *i*, this rendering of total disposition[21] is as follows:

$$D_i^{tot}(t) = V_i(t) + P_i(t) + \sum_{j \neq i} \omega_{ji}(V_j(t) + P_j(t)). \qquad [7]$$

As per [3], if at any time Agent *i*'s total disposition $D_i^{tot}(t)$ exceeds her threshold, τ_i, then $A_i = 1$ (action is taken). Otherwise, $A_i = 0$ (no action).[22] Others' thresholds do not enter into Agent *i*'s net disposition.[23]

Not Imitation of Behavior

Now, notice that in equation [7], the value of A, that is, the agent's *behavior*, is not an input. And the As of the other agents are not inputs either. No one's A appears on the right-hand side of this equation. *Hence, the mechanism of action cannot be imitation of behavior, because the binary acts of others are not registered in this calculation. So we are suspending an assumption central to the literature on social transmission.* Let us see how far we can get without it.

In principle, the sum in equation [7] extends over all humans in existence. Those not in one's network have weight zero. Socially isolated agents are, of course those with *all* non-self weights equaling zero. Throughout this book, the skeletal equation [7] alone will prove to be of interest.[24]

[21] The relationship between [6] and [7] is evident if in [6], one deletes τ_i, sets $\omega_{ii} = 1$, and multiplies out as stipulated in the preceding footnotes.

[22] None of this asserts that Agent 1 has access to V_2 and P_2. Indeed, as we will argue at some length, individuals may not even be aware of their own values. The model is flexible, however, and permits the user to explore myriad assumptions regarding any individual's imputation of disposition to others. For completeness' sake, notice again that in [7] a single weight is applied to the sum of V_j and P_j, not to either individually.

[23] In one particular extension, the Latané-Darley experiment, I have agents impute a threshold to others, because in this case the observability of others' behavior is essential to the result.

[24] Denoting the social term $[\sum_{j \neq i} \omega_{ji}(V_j(t) + P_j(t))]$ as S and suppressing time (t) for notational convenience, the skeletal disposition formula [7] is simply $D = V + P + S$: passion plus reason plus social. Mathematically punctilious readers will notice that this is different from considering V, P, and S as orthogonal bases of a dispositional vector space, a possible conceptualization mentioned earlier. This is because my model requires disposition to be a single real number comparable to a threshold, and the sum of vectors is not a real number, but an n-tuple. However, if the reader is wedded to the vector space concept—which I think is worth pursuing—one could define the basis vectors as $(V, 0, 0)$, $(0, P, 0)$, and $(0, 0, S)$. Their vector sum is (V, P, S), and my sum (D) is then simply the dot product of this with the radial vector $(1,1,1)$. So, there is a way to reconcile the basis picture with my superposition. But I have directly adopted the latter for present purposes. The vector picture invites an intriguing geometrical development, however. Staying with two dimensions, the radial vector is $(1, 1)$. Plotted in the (V, P)-plane, the radial vector represents points of equipoise

Notational Distinction

In case it is not clear, every agent in the model will be of the *Agent_Zero* type, but particular agents (instances of that agent class) will be denoted Agent 1, Agent 2, and so forth. This will be the convention throughout.

Specific Components

Now what are $V(t)$ and $P(t)$? The (emotional) functions $V_i(t)$ will be solutions of the famous Rescorla-Wagner (1972) model of conditioning.[25] The (evidentiary) functions $P_i(t)$—the bounded rationality component—will be probability estimates based on local sampling of a dynamic spatial landscape of events (stimuli).[26] And the social network is encoded in the

---

between passion and reason. If we imagine our passion-reason coordinates to be (V, P), we have from vector analysis that

$$(V,P) \cdot (1,1) = \| (V,P) \| \| (1,1) \| \cos\theta,$$

where θ is the angle between the passion-reason vector (V, P) and the radial vector $(1,1)$. It follows that

$$\theta = \cos^{-1}\left(\frac{V+P}{\sqrt{V^2+P^2}\sqrt{2}}\right).$$

But, since the numerator is just disposition, D, we may express this as

$$\theta = \cos^{-1}\left(\frac{D}{\sqrt{V^2+P^2}\sqrt{2}}\right),$$

the angle between the nonzero (V, P)-vector and equipoise. If $\theta > 0$ then $V > P$, and so forth.

[25] I will actually introduce a *nonlinear* generalization of the original model, allowing some flexibilities of interest.

[26] Net of the threshold, the skeletal equation [7] is

$$\forall i, D_i^{tot}(t) - \tau_i = V_i(t) + P_i(t) + \sum_{j \neq i} \omega_{ji} (V_j(t) + P_j(t)) - \tau_i$$

The symbol ∀ is the universal quantifier meaning "for every." Imagining just two agents, one could collect V and P terms, expressing the right-hand side as

$$(V_1 + \omega_{21} V_2) + (P_1 + \omega_{21} P_2) - \tau_1.$$

The second of these terms may, in fact, exceed unity. This might be an issue if one insisted on interpreting this term as Agent 1's probability estimate. I do not. His probability estimate is P_1, pure and simple. Other peoples' probability estimates do affect his overall disposition, but they do not affect his probability estimate proper. Now, I say, "might be a problem" because, while the probability axioms of course preclude probabilities exceeding one, innumerable psychology experiments establish

directed weights, the ω's. Again, it is *not* behavior that is contagious in these networks, but (solo) dispositions. This is just a sketch, and I will put more meat on these bones in Part I.

In the purely equation-based version of Part I, there is no spatial component, conditioning events arrive exogenously, and each agent's probability estimate is a fixed exogenous constant. Agents also do not have memory. Then, in the full agent-based version of Part II, agents roam a landscape of stochastic events, which form the basis of their immediate emotional and evidentiary states, to which network effects, memory, and further cognitive apparatus is added to produce overall action dispositions.

Coupled Trajectories

Coupled dispositional trajectories generate patterns of observable behavior in groups. Even on the emotional side alone, this is a very different picture than experimenter-subject "instructed" fear conditioning, for example. In the model developed here—and in the real world—*every agent is at once the experimenter and the subject. Each delivers stimulus to, and receives stimulus from, others, and learning is distributed, concurrent, and, often, unconscious.*

ORGANIZATION

The book is comprised of four main parts.

Part I: Mathematical Model

Part I develops the basic mathematical, explicitly equation-based, model. Relevant cognitive science is discussed as model components are developed. The most extensive discussion surrounds the neuroscience of fear

that humans make fundamental errors here. So, while I do not interpret the term of interest as a probability estimate, such an assumption could, in fact, be defensible as a model of human risk judgment. For example, suppose you ask people, What is the probability of rain today? They answer x. Now (knowing that no clouds implies no rain) you ask them, What is the probability that there are no clouds today? They give some answer y. It may well be that, from x and y, an overall probability exceeding unity is deducible. All this means is that human appraisals of likelihood violate the formal axioms of probability theory. But this is hardly controversial (See Kahneman and Tversky, 1972, 1996; Tversky & Kahneman, 1974; Dawes and Corrigan, 1974; Dawes, 1999; Fischhoff, Slovic, and Lichtenstein, 1979; and Lichtenstein, Fischhoff, and Phillips, 1982. I am not sure the term in question requires a name, since I feel no compulsion to group terms in this manner. But, forced to produce one, I might call it simply the model's "propositional output." It is the entire deliberative component, formed from the individual's P proper and a weighted sum of others' Ps proper. In any event, I do not find this to be problematic.

acquisition, which will dominate this exposition of the model's emotional component. But this is not a model of fear-driven behavior specifically. The model is general, as are the Rescorla-Wagner equations themselves, which govern many sorts of associative learning, a variety of which we shall explore. Important research in other areas, particularly social conformity effects, and cognitive biases are also discussed in the relevant sections. Part I generates the mathematical version of a key phenomenon: *the agent who initiates group action (e.g., violence) despite having no evidence, no adverse feelings, and no orders.*[27] The entire book moves back and forth between mathematical and agent-based models; this dialogue enriches both formulations.[28]

Part II: Agent-Based Model

Part II presents the second of these formulations: it is agent-based and computational. Blue individuals move about an explicit landscape of yellow patches, which we initially imagine as an indigenous population. At a stochastic rate set by the user, these yellow patches turn orange, "attacking" the Blue agents. These attacks are the conditioning trials whereby Blue agents come to associate "the Yellow face" with an aversive stimulus—an attack. This association (as in the mathematical model) is governed by the Rescorla-Wagner model. However, here (unlike the nonspatial continuous deterministic Part I version) conditioning trials occur in discrete time and stochastically as agents encounter adverse events on a spatial landscape. Fear extinction is also included in the model and is explored. To this fear-conditioning/extinction process is added the estimation, by each agent, of the probability that a random patch is an enemy (i.e., will become an orange patch). This estimate is made by *local* sampling of the landscape within a user-specified "spatial sample radius."[29] This denotes a landscape sample radius purely, and enters into the P calculation only. By contrast, agents can influence each other—have dispositional weight—at *any* range, by a large variety of avenues, including auditory, visual, and social media. The spatial sample radius is normally a set of contiguous sites on the landscape proper, such as a von Neumann neighborhood.[30]

[27] I understand an "order" to be an explicit directive from an organizational superior carrying a penalty for disobedience. There is no mechanism for orders in this model. No Eichmann "defense" is available to *Agent_Zero.*

[28] There are, in fact, two different mathematical formulations: the skeletal equations, where V and P functions are not explicitly specified, and the fully fleshed-out model, where they are. Both formulations will prove to be of interest.

[29] Occasionally, we will nickname this "vision," though the search radius must not be confused with literal ocular acuity. In general, it refers to an information set.

[30] The four immediately neighboring lattice sites to the north, south, east, and west of the agent.

This local sample yields a biased estimate of the global probability, quite in keeping with a large literature on biases and heuristics (Gilovich, Griffin, and Kahneman, 2002; Tversky and Kahneman, 1974).[31] These are the agents' *direct* V and P values, and their sum is the agent's solo disposition. But through the network, the weighted solo dispositions of others are added, and the sum (the total disposition) is compared to the individual's threshold. When total disposition exceeds that action threshold, (i.e., when $D^{net} > 0$) the Blue agent wipes out all yellow patches within some destructive radius—colored a dark blood red.[32]

Once constructed, the model is shown to generate a variety of important social parables, including our "leader," who initiates violence against yellow innocents, without evidence, without fear, without orders.

Generality

The focus on violence, like fear, is an expository tactic. I could have presented the full uninterpreted formalism and then assigned interpretations to the variables, the space, the stimuli, responses, and so forth. But this would have demanded much of the reader. So, I chose to present the model in a familiar and obviously important interpretation. However, this is not "a model of violence" per se, but of behavior in groups. And, as discussed further shortly, the general formalism admits many other interpretations. They require simply that the space, the stimulus, and the reaction be reinterpreted.

Interpretations: Vaccine Refusal, Obesity, Economic Contagion

For example, in one public health interpretation, the space could be a landscape of pharmaceuticals, orange "attacks" would be adverse drug reactions, and the destructive radius the set of drugs (e.g., vaccines) refused by fearful individuals. Vaccine refusal is a serious problem in the mitigation of contagious diseases worldwide, and (to say the least) is not uniformly driven by an informed comparison of risks and benefits. Here, the roles of emotion, partial information, and peer effects loom very large.

[31] Later we also introduce memory, so that this computation of ambush probability can be a moving average, a moving median, or some other statistic computed over a memory window. The window and sample method can vary among agents.

[32] Here, flight is precluded, *defensive aggression* (Bloom, Nelson, and Lazerson, 2001, p. 252) being the result. In the extensions flight is explored.

In an obesity interpretation, the space could be a set of foods (e.g., *x*-axis is fat; *y*-axis is carbohydrates). Orange outbursts would be opportunities for unhealthy eating, and the destructive radius is bingeing. The binge radius could be small and include just deep-fried Oreos, or it could be large and include many other foods. Eating behavior—addictive behavior generally—is an excellent candidate for the *Agent_Zero* model, since it indeed does depend on passion (the associative strength between a food's consumption and pleasure), reason (conscious deliberations regarding health), and social influence (e.g., community norms regarding diet and ideal body type).

In an economic interpretation, the space could be a set of assets (e.g., financial instruments or real estate properties), aversive events are sudden collapses in value, and the destructive radius is the set of assets dumped in response. Prices, of course, may be important in many interpretations. As demonstrated later in the book, the introduction of prices is very natural.

So, while the main exposition (particularly of the agent model) is made in terms of violence, myriad interpretations are possible. In Parts I and II, the space, the stimulus, and the action are simply reinterpreted; the core formal model is unchanged.

Part III: Extensions

In Part III the model proper is altered in a number of simple but powerful ways. These are extensions. I present the following fourteen of them:

1. Endogenous destructive radius
2. Age and impulse control
3. Fight vs. flight
4. Replicating the Latané-Darley experiment
5. Introducing memory
6. Couplings: Entanglement of passion and reason
7. Endogenous dynamics of connection strength
8. Growing the 2011 Arab Spring
9. Jury processes
10. Endogenous dynamics of network structure
11. Multiple social levels
12. *The 18th Brumaire of Agent_Zero*
13. Prices and seasonal economic cycles
14. Mutual escalation spirals

They are a heterogeneous lot designed to suggest the fertility of the approach. The fourth of these, for example, is a computational replication of the famous 1968 Latané and Darley experiment from social psychology. Here, the space is a room. The stimulus is smoke entering the room. The reaction is flight from the room.[33] Latané and Darley found that the subject's reaction is strongly influenced by the presence of nonreactive others, a result generated in a surprising way by extending the model and offering a new mechanism for bystander effects, which I term *threshold imputation*.

Entanglement

In Parts I and II, the agents' affective, cognitive, and social components have been decoupled—while disposition depends on all the components, no component is *a function of* any other. They are not entangled. In reality, they are entangled: one's emotions can directly bias one's judgments of probability. I model this in extension (6) by introducing an affective bias into the probability estimation algorithm itself.

Homophily

Also, in the discussions of Parts I and II, the interagent weights are constants. In general, these coupling strengths can vary with similarities in status, expertise, reputation, religion, musical taste, and other attributes. The *Agent_Zero* framework can accommodate any such coupling scheme. In Part III, we extend the network model in various ways. In extension (7), a dynamic affect-dependent weighting is introduced. We model the weight between two agents as a product of affective strength (the sum of their affects) and homophily (one minus the absolute value of their affective difference). This endogenous dynamic weighting then plays a central role in our models of the Arab Spring, jury processes, and the evolution of network structure.

Growing the Arab Spring

Social media enable the process of link strengthening through affective homophily. Inspired by the Arab Spring of 2011, we develop an extension where the aversive orange stimuli (the conditioning trials) are instances of regime corruption. The threshold term represents the potentially rebellious agent's perceived risk of punishment. With no social media, there is no rebellion, despite high aversion. With social media enabling amplification through affective homophily, the threshold is exceeded, and Jasmine

[33] This is why the flight extension (3) precedes Latané-Darley, for example.

Revolutions[34] unfold. This is extension (8).[35] The mutual escalation spirals of extension (14) bring to mind the violent course of events unfolding in Syria at the time of this writing (December 2012), in which indiscriminate government killing of civilians has increased participation in the rebellion, to which the government has responded with yet further civilian attacks, driving further civilians to rebel, and so forth.

Network Dynamics

Now, in classical network theory, there are simply nodes and edges. Nodes are connected to other nodes or they are not. The degree of a node is the number of other nodes to which it is connected, a completely binary matter. Link *strength* is ignored; it is zero or one. However, one could legislate that a social link exists only if the connection strength (the interagent weight) exceeds some threshold. In that case, the very structure of edges—and the degree of each node—will depend upon the underlying affinity dynamics, as well as the threshold. When the affinity curve pops above the threshold, a link is said to exist. It dissolves if the affinity returns to subthreshold level.

The classical theory might then be seen as a kind of binary projection, or embedding—akin to a Poincaré return map (see Guckenheimer and Holmes, 1983; Hale and Kocak, 1991)—of the general continuous dynamics of connection strength. This is extension (10). We compare the network dynamics on affective homophily to those on probability homophily, showing that the evolutions are very different. As in the model, so in life, we not only belong to numerous networks at once, but all of them are changing.

Jury Processes (Twelve Angry *Agent_Zeros*)

As shown in extension (9), jury dynamics offer a very rich application area, one in which affective, cognitive, and social dynamics occur in different phases. In the courtroom phase, the prosecution and defense offer data and emotional stimuli, in a battle for the hearts (V) and minds (P) of the jurors. Intrajury interactions do not (in principle) take place in this phase. So, here the weights (ωs) are zero. But then the jury is sent off to reach a verdict. All the courtroom stimuli cease (they leave the landscape of stimuli) and move to the jury deliberation room. Some jurors may have been inflamed by the evidence (high V), while others were left emotionally cold. Some

[34] So named after the Tunisian uprising, for the Tunisian national flower (Frangeul, 2011).

[35] In 2011, for example, social media facilitated the spontaneous spread of Occupy Wall Street networks and demonstrations in many countries.

found the evidence to be suggestive of a high guilty probability (P), while others did not. For some jurors, the P-value itself was amplified by passion. Some employed large memory, some very little. In these affective and cognitive states, they enter the jury phase. Behind closed doors, it may be that, as in Yeats's Second Coming, "the best lack all conviction, while the worst are full of a passionate intensity." Here weights are not zero and drive the intrajury social dynamic toward a verdict. The literature demonstrates that conformity and momentum effects play large roles here, of the sort the model generates. See Hastie, Penrod, and Pennington (1983). In the model, networks emerge through affective homophily during the deliberation phase. We model the pretrial, trial, and jury-deliberation phases of a stylized case and show how changes of venue and network effects in the jury deliberations can shape verdicts.

Economics

In extension (13) prices are explicitly introduced, and observed seasonal economic cycles are crudely generated. All in all, the range of economic applications suggested is quite wide, including financial contagion, capital flight, and various marketing strategies. Further extensions are presented below.

Replicability and Research Resources on the Princeton University Press Website

Now, talk is cheap, and so are implicit mental models. All the runs and all the extensions are explicitly implemented in *Mathematica*, in *NetLogo* Code, or both. All *Mathematica* Code is given in Appendix II. The *NetLogo* Code for the Parable 1 run is given in Appendix III. A table of all numerical assumptions used in each of the book's 14 movie runs is given in Appendix IV. All the movies, interactive *NetLogo* Applets, *NetLogo* Source Code, and a complete *Mathematica* notebook, are available at the Princeton University Press Website (http://press.princeton.edu/titles/10169.html). The Website can be used in a number of ways.

Three Ways to Use the Princeton University Press Website

1. *Play Movies.* Every movie discussed in the book is posted on the Website.
2. *Run Applets.* User-friendly *NetLogo* Applets for *every movie run* discussed in the book are also posted on the book's site. These permit nonprogrammers to manipulate sliders on the user interface and to explore different assumptions or even change them on the fly—in the course of a run. The Applets thus have both pedagogical and research value.

3. *Implement and Modify Source Code.* The *NetLogo* Source Code for every movie run is also provided in the corresponding Applet.[36] This ensures that every run is replicable and offers the modeling community a large code base to modify, extend, and use for research. The Table of Appendix IV, "Parameter Settings for Model Runs," includes all the main parameters used. These, and even finer-grained details, are, of course, also available in the *NetLogo* Source Codes themselves.

These appendices and materials on the Website ensure that all results are *replicable.* I hope the English-language exposition of the model also permits its reimplementation. But if not, the *Mathematica* and *NetLogo* programs provided are definitive. Of course, this Code Library also provides a rich basis for further extensions and explorations.

Part IV: Future Research and Conclusions

In Part IV, some ideas for future research are presented, followed by the book's overall Conclusions.

Overarching Claim

My general claim is that wherever emotional, deliberative, and social components combine to generate behavior, the *Agent_Zero* framework can apply.[37] It specializes to purely affective, purely deliberative, and purely conformist individuals. These are the "basis elements" from which a space of recognizable actors can be generated. The versatility of the framework is demonstrated with generative models from the fields of social conflict, public health, economics, law, network theory, and social psychology. I offer *Agent_Zero* as a step in the direction of a unified neurocognitively grounded foundation for generative social science.

Explanation Is Not Justification

Finally, it is worth stating explicitly that the model does not defend or justify any of the behaviors generated: violence, financial panic, or unhealthy eating. Indeed, the entire point is to offer a deeper *explanation*[38] for such dynamics, precisely in order to control them, and to better know and control ourselves.

[36] Simply scroll down from the Applet Interface to find it.

[37] A good model is like a good fugue subject: it "supports" extensive development and interpretation. So, while some may object to *Agent_Zero*, I think Bach would approve.

[38] Again, on the distinctive features of generative explanation in social science, see J. M. Epstein (2006).

PART I

Mathematical Model

IN THIS PART, we specify explicit mathematical models for the emotional, deliberative, and social components of the *Agent_Zero* framework. These choices are not cast in stone, and different components should certainly be explored, as discussed in the Future Research section. First, however, we review some underlying neuroscience of fear and its throne: the amygdala.[39]

This review is worthwhile because the Rescorla-Wagner equations (used for the affective model component) do not presuppose that fear acquisition is largely unconscious, while this is a crucially important fact from a social science standpoint, and the amygdala discussion demonstrates that it is a neuroscientifically sound modeling assumption. Also, important evidence of emotional contagion comes from fMRI studies of the amygdala, and if we didn't know anything about the amygdala, these images would mean very little.

Understanding, then, that unconscious fear acquisition is what we have in mind, we now discuss the *elementary* neuroscience of fear as prelude to the famous Rescorla-Wagner equations of conditioning, all en route to our more general model of behavior in groups.

I.1. THE PASSIONS: FEAR CONDITIONING

Humans are born with a variety of innate endowments or capacities. One of these is the capacity to acquire fear (and other) associations through a process of synaptic change in which, as Donald Hebb (1949) presciently put it, "neurons that fire together wire together." That is, after *certain*[40] pairings of an initially neutral stimulus (e.g., a tone) and a stimulus that is innately

[39] As discussed further in the context of lesion studies, the amygdala is not the only region recruited given fear-inducing stimuli. And the amygdala responds to stimuli that are not fear-inducing (Lindquist et al., 2011; LeDoux, 2012). But for the crude synthesis we have in mind, the following amygdalocentric exposition is pardonable.

[40] Not "just any" pairings will work. The exact events that lead to neuronal change and association must be stipulated with great care. A seminal article correcting many widespread misunderstandings is Rescorla (1988).

aversive (e.g., a shock), the initially neutral stimulus will evoke the same response as the innately aversive stimulus. This associative process—often termed *conditioning*[41]—is generated by synaptic change, or "plasticity." For a lucid nontechnical exposition, see LeDoux (2002). We, of course, cannot cut open a human and observe her fear, but we can intelligently speak of a fear circuit—a distributed neurochemical computational architecture[42]—whose proper functioning is of obvious evolutionary value and whose activation is strongly correlated with physical, autonomic, and other observable symptoms of fear (e.g., freezing). Indeed, LeDoux and others have mapped the fear circuit's operation in considerable detail and have made huge strides in explaining the observed capacity for associative fear acquisition, retention, and extinction by Hebbian plasticity and long-term potentiation at the cellular-synaptic level (LeDoux, 2002, pp. 79–80).

The same Hebbian picture is mirrored in the higher-level Rescorla-Wagner (RW) equations, which we shall employ in the affective component of the model. These operate not at the neuronal level but at the level of the person, or subject, where certain conditioning stimuli (the bell) become associated with specific unconditioned stimuli (the shock) through repeated pairings. There is certainly an underlying mathematical theory of neuronal function (action potentiation and firing), of which the cornerstones are the famous Hodgkin-Huxley model (Hodgkin and Huxley, 1952) and its relatives, notably the Fitzhugh-Nagamo (Fitzhugh, 1961) model. As suggested earlier, one can imagine filling in the gap between the cellular-synaptic account and the high-level RW equations with such intermediate models.[43] This is an important scientific challenge. Here, we attempt only a crude plausible synthesis of simple emotional, cognitive, and social components. But to begin at the beginning, let us examine some basic features of fear.

Fear Circuitry and the Perils of Fitness

A snake is suddenly thrown in your path. You automatically freeze. Why? From an evolutionary perspective, a reasonable hypothesis is that we freeze (are "scared stiff") because the predators faced by our evolutionary ancestors used motion detection to home in on prey, and animals (i.e., species)

[41] To be precise, it is stimulus-stimulus conditioning, as distinct from so-called operant conditioning or reinforcement learning.

[42] To some readers, this language may suggest models in the general area of "cognitive architectures," such as the SOAR model pioneered by Allen Newell. Here, I do not make use of models in this family. But see Laird, Newell, and Rosenbloom (1987).

[43] In fact, an important step would be to increase model resolution, defining the agents to be neuronal complexes in their own right. Both the computational power and the underlying neuroscience are approaching the point where exploratory modeling "from synapses to societies" is feasible. This is discussed further under Future Research.

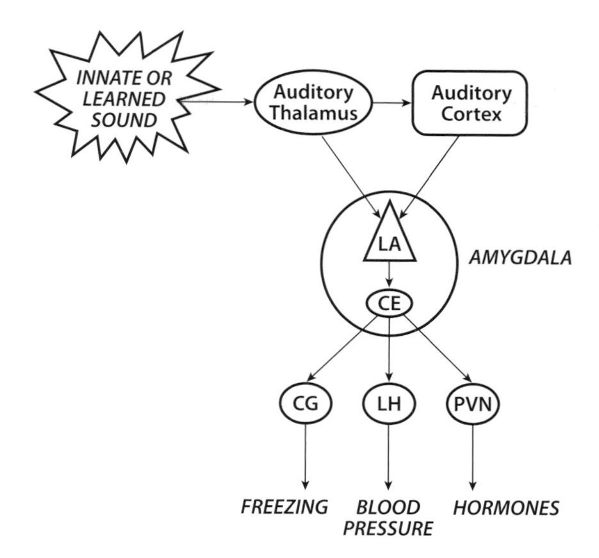

Figure 1. Auditory Amygdala Stimuli and Defense Responses. Source: LeDoux (2002, Figure 5.6)

that didn't freeze were wiped out.[44] Animals hard-wired to freeze enjoyed a selective advantage, in other words, and have passed the relevant wiring down as part of our genetic endowment.

Wiring: The Amygdala in a Nutshell[45]

As LeDoux writes, "The basic wiring plan is simple: it involves the synaptic delivery of information about the outside world to the amygdala, and the control of responses that act back on the world by synaptic outputs of the amygdala. If the amygdala detects something dangerous by its inputs (discussed further below) then its outputs are engaged. The result is freezing, changes in blood pressure and heart rate, release of hormones, and lots of other responses that are either preprogrammed ways of dealing with danger or are aspects of body physiology that support defensive behaviors." (LeDoux, 2002, pp. 8–9). A simple depiction is given in Figure 1 for an auditory threat stimulus.

Having classified an auditory stimulus as threatening (innately or through conditioning), the auditory thalamus projects (emits an action potential) to the lateral amygdala (LA) and auditory cortex, which also projects a more

[44] Relatedly, animals "play possum" to appear dead and, hence, unappealing as a meal.
[45] This is an irresistible pun, because the word "amygdala" is Latin for almond.

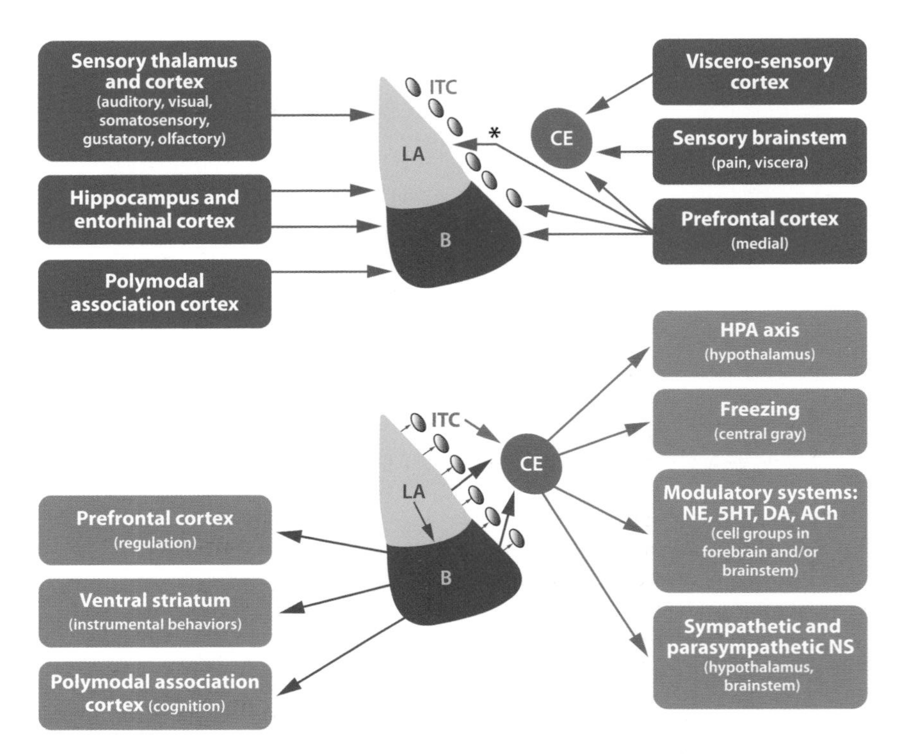

Figure 2. Amygdala Inputs and Outputs. Inputs to some specific amygdala nuclei. Asterisk (*) denotes species difference in connectivity. (Bottom) Outputs of some specific amygdala nuclei. 5HT, serotonin; Ach, acetylcholine; B, basal nucleus; CE, central nucleus; DA, dopamine; ITC, intercalated cells; LA, lateral nucleus; NE, norepinephrine; NS, nervous system. Source: Rodrigues, LeDoux, and Sapolsky (2009)

refined signal to the LA. The central amygdala (CE) then activates various systems to produce responses, such as those shown: freezing, increases in blood pressure, and the release of various hormones. (Further responses are discussed later.)

In somewhat greater detail, the neural mechanism of amygdala inputs and activation, and amygdala output, are conveyed in the diagrams of Figure 2. Inputs are depicted in the top, and outputs are shown in the bottom diagram (Figure 2).

The blue almond-shaped structure here corresponds to stunning micrographs of stained brain slices like the one shown below in Figure 3 (LeDoux, 2008).

One essential point is that this architecture supports a critical delay between unconscious and conscious responses to stimuli.

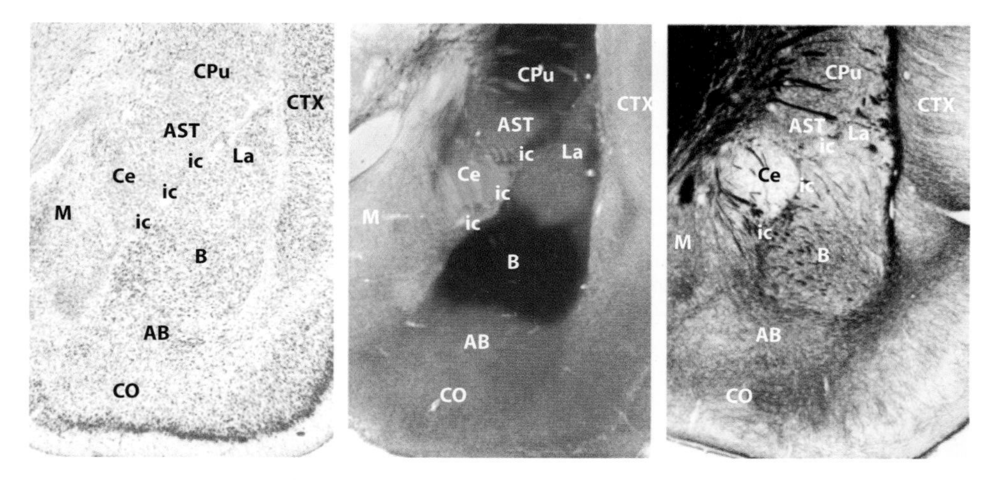

FIGURE 3. Key Areas of the Amygdala. Key areas of the amygdala, as shown in the rat brain. The same nuclei are present in primates, including humans. Different staining methods show amygdala nuclei from different perspectives. Left panel: Nissl cell body stain. Middle panel: acetylcholinesterase stain. Right panel: silver fiber stain. Abbreviations of amygdala areas: AB, accessory basal; B, basal nucleus; Ce, central nucleus; itc, intercalated cells; La, lateral nucleus; M, medial nucleus; CO, cortical nucleus. Non-amygdala areas: AST, amygdalo-striatal transition area; CPu, caudate putamen; CTX, cortex. Source: LeDoux (2008, p. 2698); reprinted courtesy of Joseph E. LeDoux

Inputs: High Road and Low Road

For example, "auditory inputs reach the lateral amygdala[46] from the auditory thalamus and auditory cortex . . . These provide a *rapid but imprecise* auditory signal to the amygdala. Cortical inputs from the auditory and other sensory systems . . . provide the amygdala with a more elaborate representation than could come from the thalamic inputs. However, because additional synaptic connections are involved, *transmission is slower*" (Ledoux, 2007). Hence, LeDoux (2002) calls these "the low road and the high road," as depicted in Figure 4.

I instantly freeze at the snake (low road) but then evaluate it as being a benign garter snake (high road), not a true black mamba, for instance. While the extreme rapidity of the unconscious response is of immense

[46] "The lateral amygdala is generally viewed as the gatekeeper of the amygdala. It is the major site for receiving inputs from sensory systems—the visual, auditory, somatosensory (including pain), olfactory, and taste systems—to this region." (LeDoux, 2007).

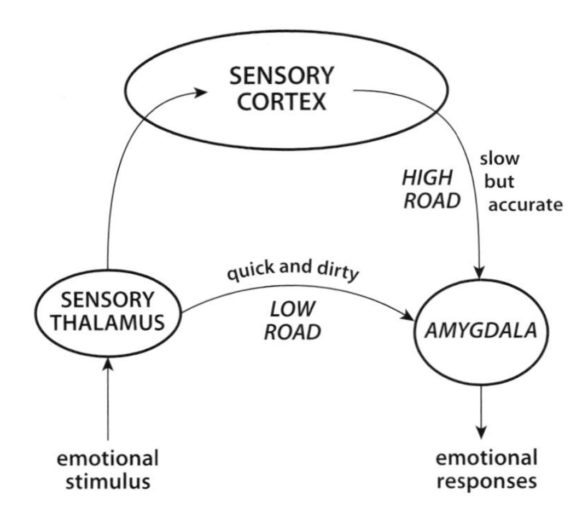

FIGURE 4. Low Road and High Road to Fear. Source: LeDoux (2002, pp. 61–63, Figure 5.7)

evolutionary value, we will see that, from a social standpoint, the lag between it and conscious appraisal is a decidedly mixed blessing.

Outputs

Continuing, "once the amygdala detects a threat, its outputs lead to the activation of a variety of target areas that control both behavioral and physiological responses designed to address the threat," (Rodrigues, LeDoux, and Sapolsky, 2009, p. 294). Beyond freezing, amygdala activation induces the release of numerous neurotransmitters (e.g., serotonin and dopamine), increasing arousal and vigilance. Endocrine and autonomic responses are also dramatic, "including increased blood pressure and heart rate, diverting stored energy to exercising muscle, and inhibiting digestion" (Rodrigues, LeDoux, and Sapolsky, 2009, p. 295).

> The pupils dilate to allow more light to enter. The heart rate picks up, and the heart muscle contracts more strongly, driving more blood to the muscles. Contractions of selected vascular channels shift blood away from the skin and intestinal organs toward the muscles and the brain. Motility of the gastrointestinal system decreases, and digestive processes slow down. The muscles along the air passages of the lungs relax, and respiratory rate increases, allowing more air to be moved in and out. Liver and fat cells are activated to furnish more glucose and fatty acids—the body's high-energy fuels—and the pancreas is instructed to release less insulin. The reduction in insulin allows the brain to draw off

> a sizeable fraction of the glucose entering the bloodstream because, unlike other organs, the brain does not require insulin in order to utilize blood glucose. The neurotransmitter that triggers all these changes is norepinephrine (Bloom, Lazerson and Nelson, 2001, p. 172).

For wonderful discussions, see also Darwin's *The Expression of Emotions in Man and Animals* (1872).[47] Contemporary scientific publications present these input-output (and additional feedback) pathways in various levels of detail. Highly detailed is LeDoux (2007).

On the experience, or "feeling," of fear, Öhman and Wiens (2003, p. 270) paraphrase LeDoux (1996):

> The fear module is primitive in the sense that it was assembled by evolutionary contingencies hundreds of millions of years ago to serve in brains with little cortices. However, it now operates in a human brain capable of advanced thought, language, and the conscious experience of emotion. Humans can talk about emotions, and they have emotional experiences. Awareness of an emotion not only depends on the recognition of an emotional stimulus but also originates primarily in feedback from the emotional responses that are elicited by the stimulus. For example, experiencing a racing heart when a shadow appears from a dark alley contributes to the feeling of fear. In fact, in perhaps the most classic of all classical contributions to the psychology of emotion, William James (1884) proposed that such feedback *is* the emotion. To paraphrase, you feel the emotion when you experience its effect on your body. Thus *the feeling of fear is the experience of an activated fear module.*

Current research on the neurophysiology of fear shows James to have been remarkably prescient, despite lacking any modern tools. Very importantly, from a social standpoint, the fear circuit can be activated, and *fear conditioning can occur, unconsciously.*

Unconscious Activation and Conditioning

In humans, "the fear module can be activated, and fear conditioning can occur without our conscious awareness." Indeed we need not *ever* become conscious of it. As LeDoux continues, ". . . unconscious operation of the

[47] His analysis of why hair stands on end, through contraction of the occipito-frontalis, is among my favorites (Ch. XII). He also records that, in cases of extreme fear, "the intestines are affected. The sphincter muscles cease to act, and no longer retain the contents of the body." The resulting name of this scared state does not require recitation here.

brain is the rule rather than the exception throughout the evolutionary history of the animal kingdom. . . . And this, moreover, confers a selective advantage . . . if we had to consciously plan every muscle contraction our brain would be so busy we would probably never end up actually taking a step or uttering a sentence" (LeDoux, 2002, p. 11). Among the many demonstrations that amygdala activation per se need not be conscious, the so-called backward masking experiments are particularly elegant.

In backward masking, "an emotionally arousing visual stimulus is flashed on a screen very briefly (a few milliseconds) and is then followed immediately by some neutral stimulus that stays on the screen for several seconds. The second stimulus blanks out the first, preventing it from entering conscious awareness (by preventing it from entering working memory)" (LeDoux 2002). But the first still elicits the full suite of physiological responses—increased heart rate, blood pressure, sweaty palms, and so forth. "Since the stimulus never reaches awareness (because it is blocked from working memory), the response must be based on the unconscious processing of the stimulus rather than on conscious experience of it. By short-circuiting the stages necessary for the stimulus to reach consciousness, the masking procedure reveals processes that go on outside of consciousness in the human brain" (LeDoux, 2002, p. 208). In short, the stimulus makes it to the amygdala by the quick and dirty "low road," but its arrival in working memory (the high road) *never* occurs. Cacioppo et al. (2007) write, "The amygdala is particularly sensitive to fear faces (Adolphs et al., 1999; Breiter et al., 1996) even when they are presented so rapidly as to not be consciously perceived (Morris, Öhman, and Dolan, 1999; Whalen et al., 1998). For another nice discussion of backward masking, see Penrose, 1999.[48]

As recent evidence of our capacity for unconscious conditioning proper, a very interesting study by Arzi et al. (2012) demonstrates that associative learning can occur even while we are asleep.

Delayed Feelings

If we do become conscious of fear-inducing stimuli, moreover, we may do so only *after* the physiological responses. Only after we have ducked from the darting bat do we notice that our heart is pounding, and we ask, "Whoa, what the heck was that!?" The conscious experience of fear, in other words, is a brain state induced by the *unconscious activation* of neurophysiological precursors driven by the amygdaloid complex (LeDoux, 2002, p. 208). Or,

[48] An important related article is Öhman and Soares (1994).

to paraphrase William James (1884), *We don't run because we fear the bear. We fear the bear because we run.*[49]

Adaptive Innate Capacity

A range of stimuli will elicit this unconscious activation—we instinctively crouch protectively at unexpected explosions nearby or when unexpected projectiles dart at our heads. In other words, certain sensory inputs will innately generate the threat response. In rats, for example, cats are in this set of innate threats. In fact, rats bred in colonies completely isolated from cats for many generations will freeze upon first exposure to cat urine (LeDoux, 2002, p. 4).

Notice, however, that animals equipped *only* with a fixed set of specific threats would be vulnerable to novel ones. So, it would be advantageous if the set could be expanded to include novel threats. And it obviously can. Pleistocene man never encountered a BMW, but *we* freeze when a car whips around the corner at us, just as *he* froze when huge animals charged suddenly from the tall brush. *We are harnessing the same innate fear-acquisition capacity—the same innate neurochemical computing architecture.* Miraculously, synaptic plasticity permits us to adapt the evolved machinery to encode novel threats. Detailed neurochemical accounts are given in LeDoux (2002, pp. 89–90).

Retention

There is little point is learning to fear hippos on Monday and then forgetting to on Tuesday. So, the *retention* of acquired fear associations is obviously essential in such cases and is achieved by various forms of long-term potentiation (LTP) at the synaptic level. This is also becoming understood

[49] James (1884, p. 190) writes, "Our natural way of thinking about these standard emotions is that the mental perception of some fact excites the mental affection called the emotion and that this latter state of mind gives rise to the bodily expression. My thesis on the contrary is that *the bodily changes follow directly the* PERCEPTION *of the exciting fact, and that our feeling of the same changes as they occur* IS *the emotion.* Common sense says, we lose our fortune, are sorry and weep; we meet a bear, are frightened and run; we are insulted by a rival, are angry and strike. The hypothesis here to be defended says that this order of sequence is incorrect, that the one mental state is not immediately induced by the other, that the bodily manifestations must first be interposed between, and that the more rational statement is that we feel sorry because we cry, angry because we strike, afraid because we tremble, and not that we cry, strike, or tremble, because we are sorry, angry, or fearful, as the case may be. Without the bodily states following on the perception, the latter would be purely cognitive in form, pale, colourless, destitute of emotional warmth. We might then see the bear, and judge it best to run, receive the insult and deem it right to strike, but we could not actually *feel* afraid or angry." [Capitalizations are in the original.]

neurochemically and is treated in Bauer, Schafe, and LeDoux (2002). Like unconscious fear acquisition, this fear retention is also significant socially, as we will discuss later in connection with "extinction."

Observational Acquisition

Finally, it would also be advantageous if one could condition on the aversive experience of others—if you could acquire fear of the red-hot stove by watching me get burned, without having to get burned yourself. As we will review, this too is possible. Indeed, so-called *mirror neurons* may have evolved for this very purpose (see the discussion on p. 62). The result is that fear is, in a defensible sense, contagious (Hatfield, Cacioppo, and Rapson, 1994). This will be further discussed shortly.

All in all, then, as LeDoux observes, "It is a wonderfully efficient way of doing things. . . ." Rather than create a separate system to encode each new danger, "just enable the [single] system that is already evolutionarily wired to detect danger to be modifiable by experience. The brain can, as a result, deal with novel dangers. . . . All it has to do is create a synaptic substitution whereby the new stimulus can enter the circuits that the pre-wired ones used" (LeDoux, 2002, pp. 6–7).

Perils of Fitness

It is indeed a most wonderful machinery. But it is also terrible: it makes us deeply vulnerable to the unconscious construction and retention of racial, ethnic, and other fears and biases [on race, LeDoux (2003); Telzer et al. (2012); on racial face masking, Öhman (2005)]. It predisposes us to rash, often violent, overreactions and opens us to all manner of nefarious manipulation. Indeed, fear conditioning has been a fundamental tool in most propaganda since time immemorial.[50] But, equally disturbing, fear can spread in a completely decentralized manner, propelling mass violence, financial crises, and deeply misguided health behaviors for example. See LeDoux (2002, p. 124):

> As Pavlov suspected, defense conditioning plays an important role in the everyday life of people and other animals. It occurs quickly (one pairing of the neutral and aversive stimulus is often sufficient) and endures (possibly for a lifetime.) These features have no doubt become part of the brain's circuitry due to the fact that an animal usually does not have the opportunity to learn about predators over the course of many experiences. If an animal is lucky enough to survive

[50] See Janis and Feshbach (1953); Higbee (1969); Pratkanis and Aronson (2001).

one dangerous encounter, its brain should store as much about the experience as possible, and this learning should not decay over time, since a predator will always be a predator. In modern life we sometimes suffer from the exquisite operation of this system, since it is difficult to get rid of this kind of conditioning once it is no longer applicable to our lives, and we sometimes become conditioned to fear things that are in fact harmless. *Evolution's wisdom sometimes comes at a cost.*" [Emphasis added.]

In other words, fitness is perilous: the innate fear module is double edged. The self-same rapid-fire, unconscious, nondeliberative fear-association machinery that allowed us to avoid predators on the African savannah leaves us profoundly vulnerable to manipulation, to unreflective acquisition of biases, and to being swept up in mass hysterias from Salem witches to genocides to the run on banks that precipitated the Great Depression.

Know Thyself

Self-knowledge (and self-control) requires that we recognize these powerfully evolved forces. Denying their existence simply increases our vulnerability to them and to their manipulation. That we possess this fear-conditioning apparatus is beyond reasonable dispute. This, however, is emphatically not to say that human behavior is *determined by* conditioning. First, the capacity to extinguish conditioned fear is also part of the innate human endowment, is also backed by overwhelming experimental evidence, and is also being mapped neurochemically. This is discussed later under the topic of *extinction*. But, beyond that, a central point of the present model is that unconscious conditioned fear may be modified both by conscious deliberation and (often unconscious) social influence. This is not to say that the overall outcome is necessarily "better than" the purely fear-inspired one; only that, as a scientific matter, (a) conditioning can be transitory and (b) much beyond conditioning is going on. Fear conditioning is incontrovertibly a part of the human condition (pun intended), and it is part of my model, along with much else.

Nomenclature of Conditioning

With all this as background, we review some standard nomenclature adopted by Pavlov in his monumental study, *Conditioned Reflexes: An Investigation of the Physiological Activity of the Cerebral Cortex.*[51] This terminology

[51] Ivan Pavlov (1927). Pages refer to the unabridged Dover edition (1960).

is necessary to present the Rescorla-Wagner model. To begin, the following are now standard definitions.[52]

Definitions

US: unconditioned stimulus [food]

UR: unconditioned response [food-induced salivation]

CS: conditioned stimulus [bell]

CR: conditioned response [bell-induced salivation]

Initialize

CS (bell) alone → 0 (no response)

US (food) alone → S (salivation)

Associative Learning

CS–US pairing trials: bell/food, b/f, b/f, . . .

When Conditioned:

B alone → S . . . CR = UR

The US is called *unconditioned* because no conditioning is required for it to elicit the response. For Pavlov's dog, food (US) induces[53] salivation without any conditioning. Hence salivation is termed the unconditioned response (UR). The conditioned stimulus (CS), by contrast, initially elicits nothing. Pavlov (1903) actually repeated his experiments with a number of different conditioning stimuli, including the famous bell. Through repeated pairings with the US, the CS acquires salience and eventually alone elicits a response, called, naturally, the conditioned response (CR). Because it emerges through repeated associations of the US and the CS, the conditioning process is also called *associative learning*.

Hume's "Association of Ideas"

Although a synaptic account of this process would require another 300 years of research, Hume recognized the general phenomenon of conditioned association, and even saw this as his signal contribution.[54] In *An Enquiry Concerning Human Understanding* (1748; 2008 ed., pp. 106–7), he writes

[52] See Pavlov (1927, p. 52). The extent and variety of experiments, the rigor with which they are recorded, and the objectivity and intelligence with which they are interpreted are exceptional. He won the 1905 Nobel Prize for his researches.

[53] Indicated by the implication symbol, →.

[54] See his "Association of Ideas." In fact, it is the cornerstone of his famous argument that there is no rational basis for belief in causation.

". . . after the constant conjunction of two objects . . . we are determined by *custom* alone to expect the one from the appearance of the other . . . Having found in many instances, that two kinds of objects—flame and heat, snow and cold—have always been conjoined together; if flame or snow be presented anew to the senses, the mind is carried by *custom* to expect heat or cold." It is not by reasoning, moreover, that we form the connection. "All these operations are a species of natural instinct, which no reasoning or process of the thought and understanding is able either to produce or to prevent" (Section V, Part I).

Hume even recognized the benefit (though perhaps not the cost) of a lag between reflexive response (low road) and deliberation (high road). He writes that, since this innate associative capacity

> is so essential to the subsistence of all human creatures, it is not probable that it could be entrusted to the fallacious deductions of our reason, which is slow in its operations; appears not in any degree in infancy;[55] and at best is, in every age and period of human life, extremely liable to error and mistake. It is more conformable to *the ordinary wisdom of nature* to secure so necessary an act of the mind, by some sort of instinct or mechanical tendency, which may be infallible in its operations, may discover itself at the first appearance of life and thought, and may be *independent of all the laboured deductions of the understanding* [emphases added] (Section V, Part I).

It is difficult to fathom so modern a perspective from someone born (in 1711) a century before Darwin (b. 1809), who would discover that Hume's "ordinary wisdom of nature" is none other than natural selection. Through the researches of Pavlov, de Cajal, Hodgkins and Huxley, and many others, we now do know something of the "mechanical tendency" Hume intuited. But it was only in the mid-20th century that mathematical models of conditioning emerge.

En route to a very famous one of these, modern nomenclature uses V for the *Associative Strength* of the CS and US. It is the extent to which the CS (the Bell) elicits the UR (salivation), or, equivalently, it is the proximity of the CR to the UR.[56] Obviously, V changes over time, with repeated pairings, and in a manner usefully captured by the Rescorla-Wagner equations (to be presented shortly).

[55] He notes that infants who experience the pain of putting their finger in a candle form an indelible association between flames and pain long before they can "reason," in his view.

[56] In other conditioning contexts, this may not be the relevant conditioning metric, as Rescorla (1988) very clearly argues.

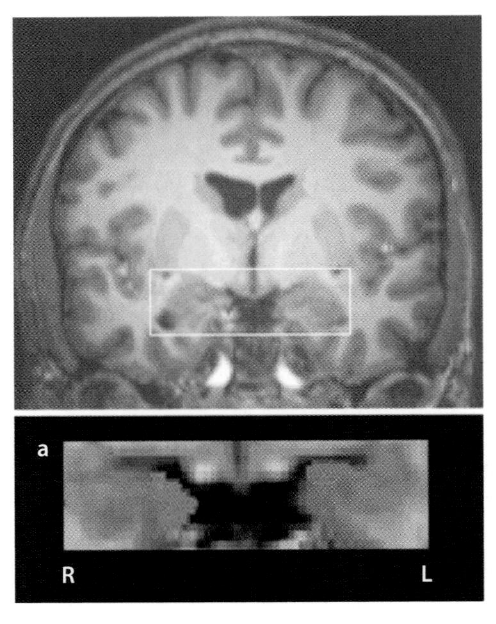

Figure 5. Postconditioning Amygdala fMRI. Source: Reprinted by permission from Macmillan Publishers Ltd: *Nature Neuroscience* (Olsson and Phelps 2007, p. 289), copyright 2007

One exemplary experiment (Olsson and Phelps, 2007) uses color as the CS, electric shock as the US, and repeated color-shock conditioning trials. After a number of these color-shock pairings, the associative strength (V) of color and shock is sufficiently great that the color (the CS) alone elicits the anticipatory shock fear, measured by conductive skin response (CSR) and by functional magnetic resonance imaging (fMRI). The fMRI image in Figure 5 shows the (blood-oxygenation-level-dependent) BOLD signal in the subject's amygdala on presentation of the CS after fear conditioning. This will prove to be of central interest below.

All brain imagery needs to be interpreted with great care (Vul et al., 2009a, b). A black-lung X-ray does not depict *the feeling of* respiratory distress and this fMRI does not depict *the feeling of* fear. But someone with a black-lung X-ray will almost certainly have trouble breathing. In medicine, feelings are symptoms. The instrumental readouts are signs. But signs are often correlated with symptoms, and that is my basic presumption here. There are many and varied correlates of fear, as reviewed earlier. Amygdala activation is a central neural one.[57]

[57]This is discussed in more detail in Part III, in connection with lesion studies.

Theory of Conditioning

There is a basic mathematical theory of the conditioning process that we shall adopt, recognizing that numerous refinements and extensions are possible. These are high-level—low-dimensional—equations developed in 1972 by Rescorla and Wagner. They do not *represent* the neural level at all. Their fidelity is *explained by* the contemporary neuroscience. They are analogous to the Kermack-McKendrick disease-transmission model (Kermack and McKendrick, 1927), which gives a very useful account of disease transmission through well-mixed populations without *representing* the microbiological interaction of pathogen and host immune systems, which, of course, *explains* transmission. From a neuroscience perspective, the Rescorla-Wagner model is a highly aggregate relationship describing an associative process generated by Hebbian plasticity at the synaptic level. It is a summary relationship explained by physicochemical synaptic interactions, which are quite well understood. So, here we are representing gross fear-learning dynamics that are explained by a lower neural level. For our social science objectives, this is a suitable modeling resolution.

The Rescorla-Wagner Model

The Rescorla-Wagner model (1972) is a cornerstone in the mathematical theory of conditioning and will form the basis for *Agent_Zero*'s emotional module.[58] Though I will generalize it slightly to accommodate a broader range of conditioning trajectories, for present purposes, Rescorla-Wagner works nicely. It is

$$v_{t+1} - v_t = \alpha\beta(\lambda - v_t). \qquad [8]$$

Exposition

Imagining a fear-conditioning exercise of the sort discussed before, let t index the paired presentations of the US and CS. So $t = 1$ is the time of first pairing; $t = 2$ is the time of second, and so forth. (One could assume that trials are a Poisson arrivals process, for example. Here, we assume

[58] Here, the term *module* refers to a software model component, not a brain region. I do not enter into the debate regarding the brain's *modularity*. To block any confusion, I typically use the term *component* when referring to model constituents. An obvious line of future research would be to swap-in alternative learning models for the Rescorla-Wagner model, such as temporal difference [Sutton and Barto (1981, p. 334), Pearce and Hall (1980), or more recent neural network models (Freeman and Skapura, 1991)]. I am not wedded to Rescorla-Wagner; I modify it in a number of ways and welcome other explorations, as discussed in the Future Research section.

these are equally spaced and identical in duration and all other respects). At each pairing there is some associative strength between the CS and the US—some degree to which the CS alone elicits (i.e., the CR approximates) the UR—for instance, the extent to which the bell alone (CS) elicits the salivation (in milliliters) normally elicited with no conditioning by food (the US). This is the value of v_t, the associative strength at trial t.[59] The pre-training association is v_0 and could be positive but will here be zero. Before any training, in other words, the bell alone elicits no salivation.

The Rescorla-Wagner model concerns the *change in* associative strength as trials proceed. The left-hand side is the *difference between* v_{t+1} and v_t. If we, advisedly, use the loaded term *learning* to denote this difference, we can coherently ask the question, When does learning stop? It stops when the left-hand side equals zero—when there is no change between v_t and v_{t+1}. The right-hand side must also equal zero, which occurs when v_t reaches the value λ, since α and β are constants (to be discussed shortly). Hence, λ represents the maximum associative strength attainable in the training process of interest and might also have been denoted v_{max}. So, if the association is already at capacity, no further gain in association is possible. Hence, the difference between λ and v_t is interpreted as *surprise* (This is sometimes referred to as the subject's prediction error). Once the associative strength of chocolate and sweetness is unity, we are not surprised when chocolate is, in fact, sweet. But our first taste of chocolate is pleasantly surprising. Finally α and β are nonnegative constants representing the salience of the CS and the salience of the US, respectively. They are often termed *learning rates*. High surprise and salience can produce very rapid conditioning. For Little Albert—recalling James Watson's infamous experiment of the 1920s—the clang of a hammer on a metal bar (the US) is salient and highly aversive, while the little furry white mouse (the CS) is initially salient and snuggly.[60] It is shocking to the infant Albert that the two would be associated so—with Watson's repeated pairings of the clang and the mouse—little Albert "learns" quickly to fear the mouse. Albert, in fact, generalized this to fear all furry animals. If either the mouse or clang had lacked salience, he might have had a pet rabbit (no aversive association would have been formed).

[59] Typically, we will use the uppercase V when the context is time independent, and the lowercase v when the context is time dependent, as in difference or differential equations.

[60] I say "infamous" because the experiment falls far short of contemporary scientific (indeed, ethical) standards. Amazingly, Watson filmed this experiment, which may be viewed at http://www.youtube.com/watch?v=Xt0ucxOrPQE. For a full account, see Watson and Rayner (2000; original appeared in 1920). Much more compelling research on conditioning has been conducted since Watson's vivid experiment, the Olsson and Phelps (2007) fMRI experiment reviewed earlier being a good example. Others are reviewed later.

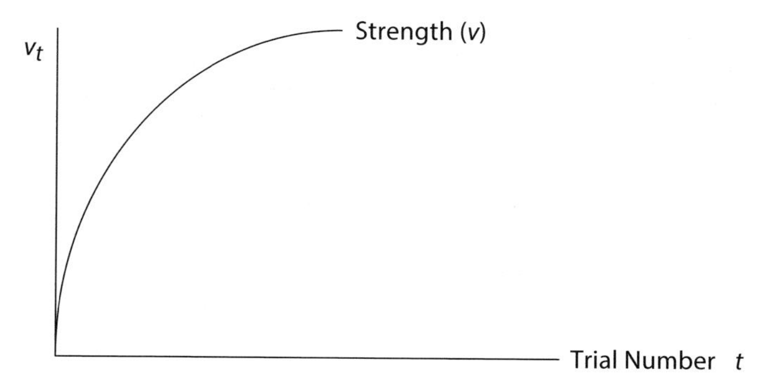

FIGURE 6. Rescorla-Wagner Associative Strength Trajectory

Observe that with α and β positive and $\lambda \geq v > 0$, v increases with each trial, but at a decreasing rate, approaching λ asymptotically. Learning—the *change in* association—is greatest at the outset, declining as the maximum association is approached,[61] as shown in Figure 6.

Unless otherwise stipulated, the variable, t, will denote trials. The Rescorla-Wagner model is a first-order initial value nonhomogeneous linear difference equation and is solvable analytically. To wit,

$$\begin{aligned} v_{t+1} - v_t &= \alpha\beta(\lambda - v_t); v_0 = 0. \\ v_{t+1} &= (1 - \alpha\beta) v_t + \alpha\beta\lambda, \end{aligned} \quad [9]$$

whose solution is:

$$v_t = \lambda[1 - (1 - \alpha\beta)^t]. \quad [10]$$

The asymptotic value of v is λ, as it should be (that is, $v_\infty = \lambda$). Now, analytical solution in hand, we can begin to ask a number of interesting questions about our disposition model.

Tipping Point

For example, assuming the preceding affective trajectory, at what trial will the individual "tip" into taking action? In the model (without deliberative or social modules yet), this occurs when $v_t > \tau$, the action threshold, as shown in Figure 7.

[61] In differential equation form, $dv/dt = \alpha\beta(1 - v)$, so the *second* derivative is $-\alpha\beta < 0$. Hence the v curve is concave down, and we see diminishing marginal associative strength.

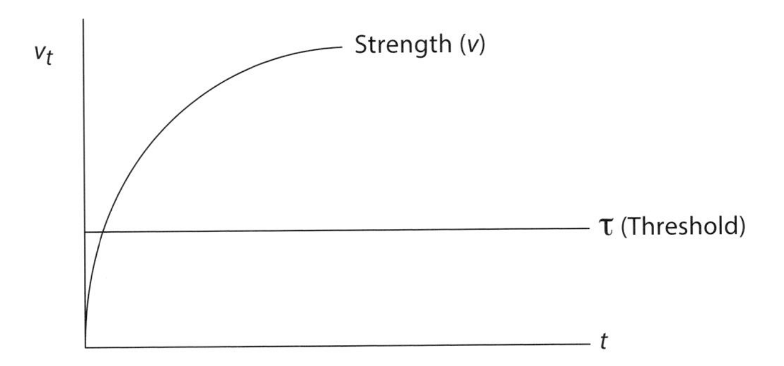

FIGURE 7. Rescorla-Wagner Trajectory and Threshold

Using the general solution, this is equivalently

$$\lambda[1 - (1 - \alpha\beta)^t] > \tau.$$

We solve for the threshold time:[62]

$$t^* > \frac{\ln\left(1 - \frac{\tau}{\lambda}\right)}{\ln(1 - \alpha\beta)}. \qquad [11]$$

This makes good sense. Increasing the threshold delays the tipping. If either α or β is zero, tipping never occurs since

$$\lim_{(\alpha\beta)\to 0} t^* = \infty,$$

meaning that if both stimuli (CS and US) lack all salience, the action is never tripped.

For completeness sake, and because we will employ it below, the differential equation (as against the difference equation) version of the Rescorla-Wagner model is

$$\frac{dv}{dt} = \alpha\beta(\lambda - v). \qquad [12]$$

[62] Starting with the first inequality, we divide through by λ and subtract 1 from both sides. Multiplying both sides by −1 reverses the inequality yielding $(1-\alpha\beta)^t < 1 - \tau/\lambda$. Now logging both sides we obtain $t\ln(1-\alpha\beta) < \ln(1-\tau/\lambda)$. Dividing through by the *negative* quantity $\ln(1-\alpha\beta)$ again reverses the inequality to yield the advertised relation (negative because $0 < 1 - \alpha\beta < 1$ and $\ln(x)$ is negative on this set).

With $v(0) = 0$, the solution is

$$v(t) = \lambda(1 - e^{-\alpha\beta t}). \quad [13]$$

The central point of the Rescorla-Wagner model, in either form, is that *high surprise combined with high salience produces strong associative conditioning.* To anticipate slightly, another distinctive feature is that conditioning depends on the aggregate stimulus, the sum of the v_i's—the associative strengths taken over all stimuli—a point to which we will return.

Once again, the model does not assume that people are *aware of* these affective dynamics. Indeed, a central point of the preceding discussion is that, typically, they are not (Phelps in Lewis, Haviland-Jones, and Barrett, 2008, p. 236; LeDoux, 2002). We may not be conscious of our conditioning (i.e., the reduction in prediction error $\lambda - v$) even if we are conscious of the CS-US pairings. And, in backward masking, even these are not registered.

Social Examples

We are well acquainted with conditioning trials in which the CS is a bell, the US is food, and the CR is salivation or where the CS is a light, the US is a shock, and the CR is fear. But, there is every reason to postulate analogous patterns of profound social consequence, as suggested in Table 1. It is surprising and profoundly salient when an unfamiliar social group attacks others, or when a trust is betrayed. Associations born of shocking and salient social events can elicit extremely strong reactions, as suggested in Table 1.[63]

Recognizing the model's generality, let us focus on one of our social examples. The September 11, 2001, World Trade Center attacks were surprising and salient. One could argue that there were four primary conditioning trials, one for each tower plus one for the Pentagon attack and one for Pennsylvania flight 93. In fact, there were countless further exposures in the form of video replays of the aircraft impacts and subsequent collapse of towers, people leaping from buildings, terrified flight, and other images. The unconditioned response (UR) was fear and intense anger toward the perpetrators. The conditioned stimulus (CS) was the face of Osama Bin Laden or Mohammed Atta—the "Arab face," as it were.

[63] In my civil violence model (J.M. Epstein, 2002, 2006), there is a computational experiment comparing (a) a large (100%) absolute reduction in legitimacy that unfolds incrementally with (b) a much smaller (30%) absolute reduction, but in which the drop occurs in one step, not incrementally. The latter produces a much stronger public reaction in the model, consistent with Rescorla-Wagner.

Table 1. Surprise, Salience, and Conditioning

CS	US	UR/CR
Light	Shock	Fear
Japanese face	Pearl Harbor	Rage/internment
Arab face	9/11	Rage/internment
Vaccine	Report of adverse reaction	Fear/vaccine refusal
Doctor	Tuskegee	Enduring distrust
Asset	Collapse	Dumping

The resulting associative strength was extremely high, as expected on good Rescorla-Wagner (RW) grounds.[64] Most Americans had little prior exposure to Muslims and certainly had never heard the phrase "Al Qaeda" before, so there was no damping of the association by prior conditioning. And, as expected, we saw rapid "learning," in the RW sense. After the four direct trials and countless reexposures in all media, the average CR to symbols of the Muslim world (CSs) was very high up the learning curve.

"A comprehensive LexisNexis database survey of U.S. newspaper reports between September 1 and October 11, 2001, found an increase in hate crimes toward persons believed to be of Middle Eastern descent (from 1 to 100 events involving 128 victims and 171 perpetrators) across 26 states" (Swahn et al., 2003; Marshall et al., 2007, p. 311). Fourteen of these were murders. "Most [attacks] occurred within the period 10 days after the 9/11 attacks" (p. 311).

Remarkably, "only 42% of the victims were of Middle Eastern descent," the remaining attacks being "against persons of color who are perceived to be vaguely reminiscent of the 9/11 terrorists" (Marshall et al., p. 311). Even a very broad and vague attribute (general skin hue) can serve as a CS. This is an example of stimulus overgeneralization, where subjects conditioned on a particular CS—an 800-hz tone—will respond to a very rough approximation of it (e.g., a 1000-hz tone).

Olsson and Öhman (2009, p. 736) write, "For example, *there are now numerous demonstrations that unknown racial outgroup members, that is, individuals not belonging to one's own racial group*, can elicit a rapid threat response associated with the amygdala" (Cunningham et al., 2004; Phelps et al., 2000). They even speak of "the possibility of a hard-wired disposition to develop xenophobic responses" (p. 736). This is not to say that xenophobia itself is inevitable. Indeed, they also note that out-group dating experiences can nullify the effect. Why this in-group bias might have evolved is modeled

[64]This example comes from a talk John Cacioppo delivered at the Social and Behavioral Workshop, sponsored by the National Insitute for General Medical Sciences in Washington, DC, in November 2008.

in Hammond and Axelrod (2006a, b). The first fMRI study of prejudice was Hart et al. (2000; see also Cunningham & Van Bavel, 2009, p. 978).

Blocking and Selective Discrimination

Relatedly, I find it very revealing that the Japanese were the only ethnic group interned on a mass scale by the United States during WWII, even after 1944, when the United States was fully at war with Germany and Italy. In 1939, The German-American Bund had staged a 20,000-person pro-Nazi rally in New York's Madison Square Garden. The Bund ran a dozen youth camps in various states and published eight newspapers. Once the United States entered the war, the Bund was banned, but, unlike the Japanese (who had never organized anything comparable), few German-Americans were interned. Even fewer Italian Americans were interned, despite fascist Italy's alliance with Hitler. Beyond the unique scale of their internment, the Japanese were distinctive in that more than 60% of those interned were, in fact, American citizens.[65]

This, too, is entirely consistent with the general version of the Rescorla-Wagner model, in which the total associative strength v^{TOT} is distributed over the *sum of* all conditioning stimuli of relevance. If we let v_t^J stand for the associative load on the Japanese at time *t*, and v_t^E the load on European axis powers, the total strength is given by[66]

$$v_{t+1}^{\mathrm{TOT}} - v_t^{\mathrm{TOT}} = \alpha\beta\,[\lambda - (v_t^E - v_t^J)]. \qquad [14]$$

If the associative strength of v_t^J is already close to λ, there is very little associative capacity left for v_t^E. In these terms, the associative load on the Japanese face v_t^J was so large after Pearl Harbor (shocking and salient) and the ensuing war in the Pacific as to "block" a comparable association on Aryan features or Italian accents.

Interestingly, very few Japanese *in Hawaii* were interned.[67] It could be that they were grudgingly tolerated as essential to American naval base operations. But one could also explain this by Rescorla-Wagner: Japanese people were part of the fabric of Hawaiian society, composing a third of

[65] This overarching point being true, the precise details of internment policy toward specific subpopulations (e.g., legal citizen vs. alien) of each group is quite tangled. See, for example: Semi-annual Report of the War Relocation Authority for the period January 1 to June 30, 1946, not dated. Papers of Dillon S. Myer. Scanned image at trumanlibrary.org. Retrieved December, 2012.

[66] For the full matrix version of the RW model, see Danks (2003).

[67] I thank Robert Axelrod for this salient contrast. "All Japanese who lived on the West Coast of the United States were interned, while in Hawaii, where the 150,000-plus Japanese Americans composed over one-third of the population, an estimated 1,200 to 1,800 were interned," http://en.wikipedia.org/wiki/Japanese_American_internment; *Japanese Americans, from Relocation to Redress.* 1991, page 135.

the population. Non–Japanese Americans living in Hawaii had accumulated sufficient positive experience (prior exposure) as to "block" the level of fear that continental Americans associated with the (far less familiar) Japanese after Pearl Harbor. Analogously, on 9/11, most Americans had no such prior exposure to Muslims, and no blocking of the maximal associative load occurred. The phenomenon of blocking has been studied extensively, beginning with the classic paper of Kamin (1969).

Betrayals Real and Imagined

Betrayals of trust are often very surprising and salient. The Rescorla-Wagner model may explain why they generally loom so large in human memory. The betrayal of trusting black Americans by the medical establishment at Tuskegee is a stark example. It was very surprising and highly salient. It actually continued until 1972 and so is a deep trauma well within the memory of black Americans today. Judas betrays Christ; Brutus betrays Caesar; Greek mythology is rife with betrayals (Clytemnestra betrays Agamemnon); "Uncle" Joe Stalin (ally in WWII) betrays the war allies by occupying Eastern Europe; the Jews allegedly "stabbed Germany in the back" after WWI.[68] Benedict Arnold betrays the colonies. They are instances of highly salient surprise. The "revelation" of betrayal by conspiracies is a trusted tactic among fear mongers to this day. See Richard Hofstadter's (1964) wonderful essay, "The Paranoid Style in American Politics."[69]

By the same token, some salient surprises are reserved for occasions meant to elicit a burst of strong and happy associative strength—like marriage proposals. Many expectant parents choose not to learn the sex of their children till birth, preserving a highly salient surprise.

So, here we have our first building block of *Agent_Zero*—Rescorla and Wagner's very elegant model of conditioning. We understand that this is an aggregate relation that is ultimately explained by the neuroscience, which *licenses us to interpret the model* as unconscious fear acquisition harnessing the same Pleistocene apparatus that got us here,[70] a neurophysical apparatus that was reviewed in some detail.

Now, as noted, a number of factors can militate against our blind submission to unconscious fear impulses. Counterevidence is one; peer pressure is another. These will both be further building blocks added to the model.

[68] This was a common anti-Semitic myth in the interwar period in Austria and Germany.

[69] Joseph McCarthy claimed to have revealed "a conspiracy on a scale so immense as to dwarf any previous such venture in the history of man." Hofstadter (1964). As Hofstadter writes, "The paranoid spokesman sees the fate of conspiracy in apocalyptic terms—he traffics in the birth and death of whole worlds, whole political orders, whole systems of human values." The images should be surprising and salient on a world scale.

[70] Hebbian learning presumably evolved long before humans did.

But, even within the Rescorla-Wagner framework, there is allowance for the fading of fear.

Fear Extinction

Our synapses are plastic, and so are we—we can learn, and we can unlearn (Rescorla and Wagner, 1972). Conditioned associations are not necessarily permanent and often decay if pairing trails cease. This is called *extinction,*[71] a term introduced by Pavlov. In the Rescorla-Wagner (RW) model, the extinction phase is handled very elegantly simply by setting λ, the limiting value of v, to zero (since all association is to die out), and the initial value of v to whatever value it had attained immediately when conditioning trials are terminated; let us denote this latter value v_{max}. In differential equation form (moving freely between discrete and continuous time), associative strength thus diminishes according to

$$\frac{dv}{dt} = \alpha\beta(0 - v), \text{ with } v(0) = v_{max}. \qquad [15]$$

The solution is the well-known formula for exponential decay,

$$v(t) = v_0 e^{-\alpha\beta t} = v_{max} e^{-\alpha\beta t}. \qquad [16]$$

Overall, then, the conditioning and extinction phases of an RW process are *not* symmetrical, and most likely involve different brain regions [as discussed shortly]. Conditioning is increasing and concave down, with an upper asymptote of λ. Extinction is decreasing and concave up, with an asymptote of zero. When conjoined the acquisition and extinction phases have a distinctive shape, with an abrupt change in concavity at the (nondifferentiable) acquisition-extinction transition point, as shown in Figure 8.[72]

General Solution of the Two-Phased Model

Typically, the two phases (acquisition and extinction) are solved and discussed separately. I have not seen it observed that the entire two-phased model can be expressed using Heaviside unit step functions.[73] With t^* the time at which trials cease, the full solution is then

$$v(t) = \lambda(H(t^* - t)(1 - e^{-\alpha\beta t}) + (1 - e^{-\alpha\beta t^*})H(t - t^*)e^{-\alpha\beta(t - t^*)}). \qquad [17]$$

[71] Even the experience of phantom limbs can eventually fade. Neurons with active function will literally displace those formerly engaged in the regulation of now-amputated limbs.

[72] Figure 15 was drawn to reflect these properties.

[73] Again, we specifically define: $H(x - y) = 0$ if $x \leq y$ and 1 if $x > y$.

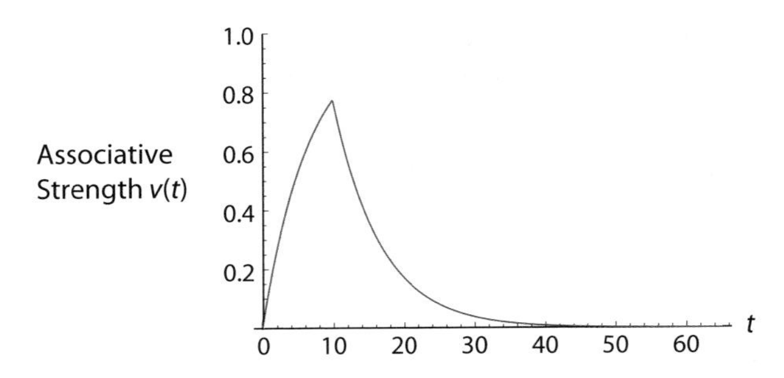

FIGURE 8. Acquisition and Extinction

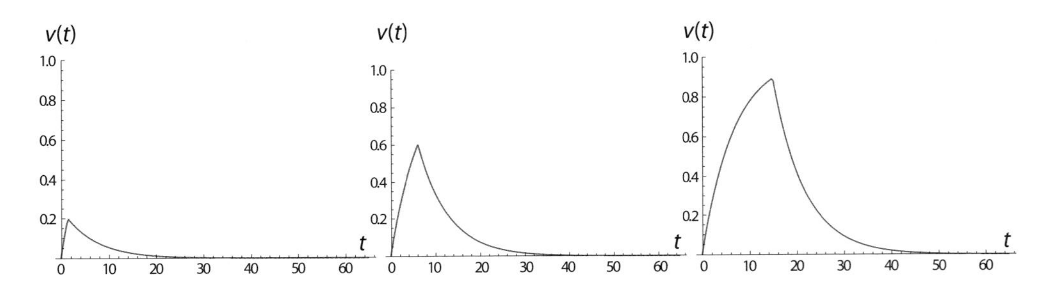

FIGURE 9. Final Acquisition Level and Initial Extinction Rate

This gives the entire learning and extinction trajectory. A movie of the entire acquisition and extinction history, as parameters are varied, is given as **Animation 0** on the book's Princeton University Press Website.[74]

An impressive property of the two-phased model is that the (negative) extinction *slope* increases in magnitude with the terminal phase 1 conditioning *level*, as noted in the panels of Figure 9. The greater the acquisition *level*, the greater (i.e., steeper) the initial extinction *rate*.[75]

The curious asymmetry of the model makes its extensive corroboration all the more impressive. Countless experiments with animals, including humans, have conformed to this basic relationship. For example, conditioning trials with humans exhibit the same qualitative profile as the following rat trials (Figure 10).

[74]The *Mathematica* movie code from which these screen shots were taken is Animate[Plot[UnitStep[n − t](1 − Exp[−.15t]) + UnitStep[t − n](1 − Exp[−.15n])Exp[−.15(t − n)], {t, 0, 65}, PlotRange → {0, 1}], {n, 0, 30, .1}] also given in Appendix II.

[75]Alternatively, the closer to horizontal is the acquisition curve, the closer to vertical is the extinction curve.

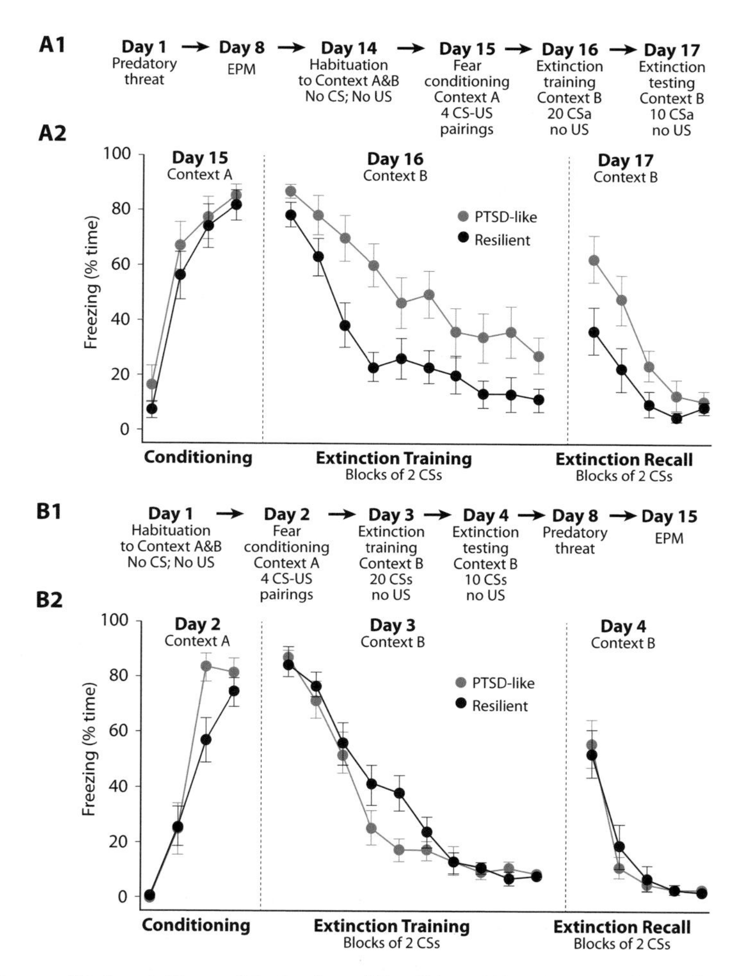

Figure 10. Acquisition and Extinction of Conditioned Fear Response for Predatory Threat. Source: Goswami et al (2010, p. 496)

Moreover, there are strong arguments as to why the animal models are reasonable predictors of human fear conditioning behavior (Bloom, Lazerson, and Nelson, 2001). Obviously, we do not fear *what* the rat fears, but we fear *how* the rat fears.[76] The same neuroanatomy and cellular-synaptic mechanisms have been preserved across vertebrate evolution. LeDoux calls these conserved structures "survival circuits." In his most recent

[76] I am speaking somewhat figuratively here.

terminology, the fear circuitry we've reviewed is cast as an instantiation of these. For a full account see LeDoux (2012). A compendious review of the human conditioning literature is Sehlmeyer et al. (2009).

Although the Rescorla-Wagner model has been refined in Pearce and Hall (1980), Sutton and Barto (1998), and other descendants, the classic model deserves the status of other canonical models—the Kermack-McKendrick model of infectious disease; the Lotka-Volterra model of predator-prey cycles, the Richardson arms-race model, and so forth. Like fundamental models in many fields, it elegantly offers important insight and explains a wide range of observed phenomena.[77] It is a revealing simple model.

Affective Persistence: The Half-Life of Hatred

Because extinction is seldom immediate, affect (positive or negative) can remain above the action threshold long after the stimulus has stopped. Cycling back to our social examples, then, anti-Japanese sentiment generally continued beyond the war. The informal Jewish boycott of German goods persisted (indeed persists) long after WWII. As public health examples, the scars of Tuskegee still affect minority trust of the U.S. public health establishment (Corbie-Smith, Thomas, and St. George, 2002; Freimuth et al., 2001). Such distrust is evident in The Pittsburgh Barbershop study (Using Social Norms to Attack Prostate Cancer among African Americans, National Center on Minority Health and Health Disparities), in survey results on smallpox vaccine refusal (Lasker, 2004), and in the Washington, D.C. postal workers' cipro (ciprofloxacin) refusal after the anthrax attacks of 2001(Quinn, Thomas, and Kumar, 2008; Quinn, Thomas, and McAllister, 2005). The last example is stark in that predominantly white Congressional staff were eager for cipro. The same general pattern occurred with H1N1 (swine flu) vaccine in 2009, despite the swine flu being declared a global pandemic by the World Health Organization.

Fear conditioning and the extinction of fear, in other words, are not symmetrical, a point made nicely by the Rescorla-Wagner model. It is amazing that seemingly remote processes can share the same mathematical description [e.g., the wave equation; see J. M. Epstein (1997)]. Even in social science, a huge number of social situations have the form of a Prisoners' Dilemma, or a Coordination Game.[78] Here also, the extinction phase of

[77] On the value of simple idealized models, see J. M. Epstein, (2008).

[78] Similarly, so-called gravity models of economic interaction among cities posit that the force of attraction between agglomerations is proportional to the product of their sizes and is inversely proportional to a power of the distance between them, as in Newton's law of universal gravitation. "Size," of course, can be interpreted variously, and "distance" can be replaced with transportation cost, or effort required, none of which changes the gravity model parallel, for which I thank Robert Axelrod.

Rescorla-Wagner is formally the same as radioactive decay. So, just as we could compute the tipping point earlier, let us compute the "half-life" of hatred, if you will. The half-life is, by definition, the time at which half the original "substance," v_{max}, is gone. It is the time at which

$$\frac{v_{max}}{2} = v_{max}e^{-\alpha\beta t}. \qquad [18]$$

An interesting property of exponential decay is that the half-life is independent of the initial level, as in this case, where the v_{max}'s cancel out. Accordingly, logging both sides, we obtain

$$\ln\left(\frac{1}{2}\right) = -\alpha\beta t,$$

which is to say that the half-life is

$$t_{half} = \frac{\ln(2)}{\alpha\beta}. \qquad [19]$$

This makes basic sense in that the smaller the decay rate ($\alpha\beta$), the greater the half-life (i.e., the longer it takes until half the initial stuff is gone).

Posttraumatic Persistence

Of course, the mere cessation of conditioning trials is not always sufficient to "reset lambda to zero" and induce the exponential extinction of fear. As LeDoux and Phelps write, "It is important to note . . . that the extinction of conditioned fear responses is not a passive forgetting of the CS-US association, but an active process, often involving a new learning" (LeDoux and Phelps, 2008, p. 164). In fact, acquisition and extinction of fear may be controlled by different regions—acquisition by the amygdala, and extinction by the anterior cingulate of the medial prefrontal cortex (mPFC). A 2005 PET study of women with PTSD resulting from childhood sexual abuse revealed "decreased function or failure of activation in mPFC during fear extinction, in women with abuse-related PTSD compared with controls" (Bremner et al., 2005).

Again, we are not modeling brain regions, but in modeling terms, we have license to say that λ doesn't necessarily reset to zero when conditioning trials cease (simply because genocidal violence stops, for example).[79] Below,

[79] Once frightened by a snake, if you avoid all snakes, you never experience a harmless snake, and so you never lose your fear of all snakes. In the larger sociopolitical setting, the same inertia (never talking to members of a demonized out-group) can prolong intergroup hostility.

we exercise this license mathematically (see Figure 33) and show that a single agent's PTSD can retard the recovery of the entire group. Then, in the agent-based model of Part II, we (I believe for the first time) "lesion" an agent—excising her "software amygdala"—and show the group-level effects.

The Rescorla-Wagner model will be generalized in several ways below. But it will form the backbone of *Agent_Zero*'s affective component. We turn now to the cognitive (evidentiary/deliberative/ratiocinative) building block of *Agent_Zero*, having agreed with Hume and countless others, that reason (not only passion) must play a role in any credible model of people.[80]

I.2. REASON: THE COGNITIVE COMPONENT

However, reason is not here assumed to be perfect, but prone to informational limits and associated biases. There is, of course, a vast literature on *bounded rationality* since Herbert Simon coined the phrase (see Simon, 1982). One can imagine endowing agents with innumerable sources of error. Well-established and systematic departures from canonical rationality include: representativeness and availability biases, anchoring and adjustment, recency effects, the conjunction fallacy, confusion between frequency and magnitude, base-rate neglect, and outright logical confusions, to name a few (see Gilovich, Griffin, and Kahneman, 2002; Kahneman, 2003). To start, however, I need the agents to estimate a probability. Why will this be important? Here is the motivation, as always in the context of our binary disposition model.

Imagine that I am confronted with another person, Mr. K, and the binary issue is whether to attack him or not. And, as per our earlier discussion, with *no* countervailing forces (e.g., fear of retaliation), I will fight K if my visceral *feeling* (V) that he is my enemy exceeds my action threshold T. If I feel nothing toward K and the *data* on him indicate a probability (P) of enmity below T, then I will not attack. Suppose now that my feelings (passion) exceed threshold but that my evidence (reason) does not. Then the emotional and cognitive circuits are giving me conflicting signals.[81] Since, tautologically, I either attack or do not, one of these prevails. Exactly how this happens is not well understood but normally comes under the heading of "executive function," in which certain regions of the prefrontal cortex (PFC) are centrally implicated. We will return to this.

But, to mathematize this in the simplest conceivable way, the situation just described is depicted in Figure 11.

[80] This admonition will certainly have a strange ring to it for those raised in the formal rational choice tradition, where it is *affective* dynamics that play virtually no role.

[81] Some would call this cognitive dissonance, á la Festinger (1957).

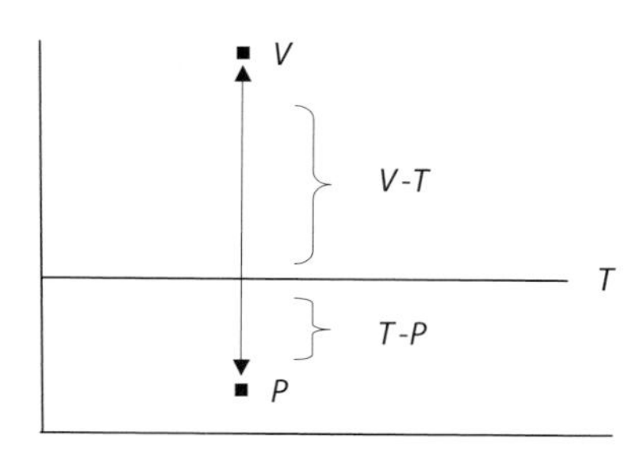

FIGURE 11. Rivalry

The action threshold is T. Emotion (V) is pulling me upward, urging me to attack. But the evidence (P) suggests I should not. Of course, the action in question could be to pursue an amorous inclination, rather than a violent one, or to flee rather than fight, and so forth. A more colorful rendition of the basic tension is Plato's allegory of the Charioteer.[82]

In *Phaedrus*, Plato depicts man as a Charioteer with two horses, an upward-striving white horse who can lift the Charioteer to the heavenly, reasoned, temperate altitude of the Gods, and a downward-plunging black horse whose wanton animal impulses, if unbridled, will bring the Charioteer crashing to earth.

The black steed can be controlled, but only through great and violent effort. In such moments, the Charioteer, yanking the reins, "falls back like a racer at the barrier, and with a . . . violent wrench drags the bit out of the teeth of the wild steed and covers his abusive tongue and jaws with blood, and forces his legs and haunches to the ground and punishes him sorely. And when this has happened several times and the villain has ceased from his wanton way, he is tamed and humbled, and follows the will of the Charioteer" (Edman, ed., 1956, p. 297). For Plato, one steed strives upward, and one strives downward. They pull in opposite directions on the same line, as in a tug of war, in contrast to the orthogonalist Hume. Of course, Hobbes accounted the Charioteer's prospects as grim, arguing that a monolithic Leviathan state is required to bridle, as it were, our warlike impulses, lest 'the life [of man] be solitary, poor, nasty, brutish, and short" (Hobbes, 1660; 1958 ed.).

Returning to the mathematics, in Figure 11, passion (V) pulls up, but evidence (P) pulls down.[83] Oddly, my model says to *add* them (see equation [1]). Why? Returning to our first example, the argument is that, to attack, V must be pulling up more than P is pulling down—in other words, that V

[82] I thank Duncan Foley for suggesting this.

[83] This P should not be confused with passion, although, admittedly, it would be a natural variable name.

exceeds the threshold by more than the threshold exceeds P, as shown in Figure 11.

Mathematically, this means $V - T > T - P$, or that

$$V + P > 2T. \qquad [20]$$

Passion *plus* reason must exceed twice the threshold, as it were.[84] So, addition—first introduced in the skeletal equation—is a defensible starting point, but as we see, a further argument can be provided. The sum, moreover, must satisfy a perhaps unexpected relation. Another way to think of this, of course, is that the average, $(V + P)/2$, must exceed the threshold. Obviously, if V and P are both above (or both below) T there is no rivalry, while in Figure 11, no action is taken (i.e., reason wins) if $(V + P)/2 \leq T$.[85]

As reviewed before, the Rescorla-Wagner model produces a *V-curve*, not just the point in Figure 11. So, assuming that $P < T$, the full dynamic picture is shown in Figure 12. Instead of the *point*, we now imagine associative strength developing through trials as per the Rescorla-Wagner model. The action threshold remains $V + P = 2T$, but V is now a curve, not a point. If and when V exceeds $2T - P$, *Agent_Zero* acts (e.g., attacks).

All right, so this is how *Agent_Zero*'s probability estimate, P, will *function* in the model; this is how he *uses P*. But where does *Agent_Zero* get a value for P?

In the agent-based version of Part II, he computes it from data he collects. The agent model developed there is spatial. Events unfold on a grid of cells. Agents have a *spatial sample radius*. If this sample radius is 1, they can survey the adjacent cells immediately to the north, south, east, and west, their so-called Von Neumann neighborhood.[86] If the radius is 2, they can inspect eight cells (two in each of the four directions), and so forth. They collect data through observation of cells within their search radius. For example, if we imagine neighboring cells to be people, and posit that yellow cells are friends and orange cells are enemies, then *Agent_Zero*'s estimate of the global probability that a randomly chosen person is an enemy is his

[84] Attentive readers will notice a tiny discrepancy with the action rule [3] of the Introduction. Using the earlier-developed terminology, solo disposition $(V + P)$ would have to exceed *twice* the threshold, rather than simply the threshold. The reader may feel free to define the τ of Action Rule [3] as $2T$, as defined here. But, in all candor, since we are not performing numerical calibration, I see no harm in ignoring it for notational convenience, and henceforth will.

[85] The reader is invited to check that this obtains regardless of which variable, V or P, is playing the role of action proponent.

[86] The spatial search radius is a user-adjustable slider (global variable) in the *NetLogo* Interface. *NetLogo* permits the user to toggle freely between Von Neumann and Moore neighborhoods. The latter includes the diagonal—NE, NW, SE, and SW—neighboring sites, for a total of eight for a radius of 1.

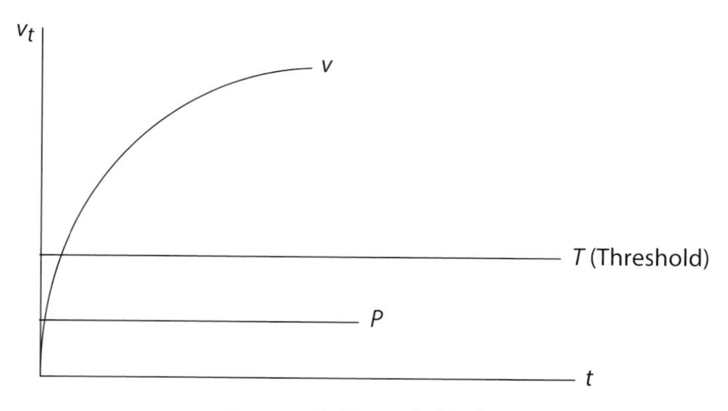

FIGURE 12. Dynamic Rivalry

local (i.e., within radius) relative frequency of orange sites. We will interpret this information radius in a variety of ways later (e.g., size of an investment portfolio, range of vaccines). But one interpretation is the literal sensory perception of events in space and time.

Sample Selection Error

In general, agents want to know the prevalence of some attribute (orangeness) in the *overall* population of orange and yellow cells. They estimate this by computing the *local relative frequency of* orange cells within their spatial sample radius, as discussed earlier. This algorithm produces bias because this color may be clustered in certain regions. Compared to the true global orange frequency, sampling in high-density zones produces upward bias, while sampling in denuded areas produces downward bias. Statistically, this is referred to as *sample selection error*. Cognitively, it arises from what Tversky and Kahneman (1971) dubbed the *representativeness heuristic*: the common tendency to "expect the statistics of a sample to closely resemble (or 'represent') the corresponding population parameters, even when the sample is small" (see Tversky and Kahneman, 1971, and Kahneman and Frederick, 2002).

Memory

Now, just as agents can look around in space, they can also look back in time. In the agent-based version of the model, they will have "memory"—they will maintain a list—of recent probability estimates (relative frequencies within their spatial sampling radius). If they have memory m, they carry a list of the m most recent values (dropping memories from more than m periods ago). While the code permits various filters (e.g., the moving median), agents will use the *moving average of probability estimates over this memory window* as their P value. While they are not Bayesians, they do update their estimates based on new data, and prior estimates have inertia. The essential point is that *the agents in this model do not remember everything, and what they do remember is typically biased.*

Obviously, human memory is a vast and dynamic field of cognitive neuroscience in its own right (Eichenbaum, 2012) and one over which I make no claim to mastery. *Agent-Zero*'s idealized memory (her evolving list) will serve as a simple starting point in building an agent who can take in and store information from a dynamic environment and estimate from it a probability that affects disposition and, in turn, action.

Even these seemingly crude modeling assumptions, it should be noted, impute to *Agent_Zero* considerably more numerical prowess than humans can muster without serious training. Butterworth (1999) argues that the computation of frequencies is very difficult for humans and that base rates are hard to estimate because rates per se are hard, hence suggesting a neurocognitive basis for the robustly observed behavioral regularity of base rate neglect (Kahneman & Tversky, 1973).[87] Moreover, the storage of numbers beyond subitizing (i.e., remembering four values) is also difficult. We will return to the topics of sample selection bias and memory in the agent model of Part II.

Probability in Agent vs. Mathematical Versions

Summarizing the probability discussion, in the agent version developed shortly, the agents' P-value (a) is acquired through spatial sampling, (b) is remembered, and (c) is dynamic (i.e., it updates) as the agents' environment changes and agents move spatially. In short, it changes based on the agent's experience. For expository purposes, in this purely mathematical section, it will be *exogenous and fixed.*

So, we now have simple versions of two of our three *Agent_Zero* ingredients: We have a crude model of affective dynamics (passion) using the Rescorla-Wagner Model (to be generalized later to accommodate S-curve learning); and we have a crude model of bounded rationality—our

[87] Base-rate neglect precludes proper Bayesian updating, because conditional probabilities are computed incorrectly. Bayes' theorem is that: $P(x|y) = P(y|x)[P(x)/P(y)]$. The base rate refers to the bracketed term. Neglecting it leaves the so-called base rate fallacy that $P(x|y) = P(y|x)$. This is quite a common, and very important, mistake. The probability of someone being a Muslim given that they are an Islamic terrorist, $P(M|T)$, differs vastly from the probability that someone is a terrorist given simply that they are Muslim, $P(T|M)$. Base-rate neglect equates the two, often producing baseless suspicion or worse. This is not to say that no people apply Bayes' rule properly. But our agents use a different algorithm. An extension mentioned later is that agents could be Bayesians. The population might also be heterogeneous, with a mix of Bayesians and other inferential species.

P-estimate (to be generalized and made dynamic and spatial in the agent-based version to come).

Let us now imagine again that the wriggling snake is thrown in our path. First we have the "low-road" unconscious response reviewed earlier: I freeze in primal terror. But then I notice that this could be a simple disinterested garter snake. I assign some P to that prospect. My ultimate behavior in the snake's presence—flee or just cut a wide swath and pass by—is neither purely emotional nor purely rational. There is a competition or rivalry between them, in which rationality often operates at a distinct disadvantage. Indeed, Charles Darwin himself was deeply interested in this specific example. He actually tried "without success, to withhold a response to the strike of a poisonous snake behind a protective glass cage in a zoo" (LeDoux, 2002, p. 216).

Recap and Transition

To this point, we have developed two of *Agent_Zero*'s three components: one affective and one deliberative. But "no man is an island," as it were, and it is now time to add the social component. Elaborating our snake story, first we freeze (the emotional) and then we calm down, seeing that the snake is likely benign (the cognitive). But, if a crowd of horrified people race by shrieking "snake," we may abandon our cognitive appraisal and run (the social). We now add a social component.

I.3. THE SOCIAL COMPONENT

The simplest conceivable society is comprised of two people. In Figure 13, the vertical axis is person 1's disposition, and the horizontal axis is person 2's. Each has an action threshold, τ_1 and τ_2, respectively, dividing the plane

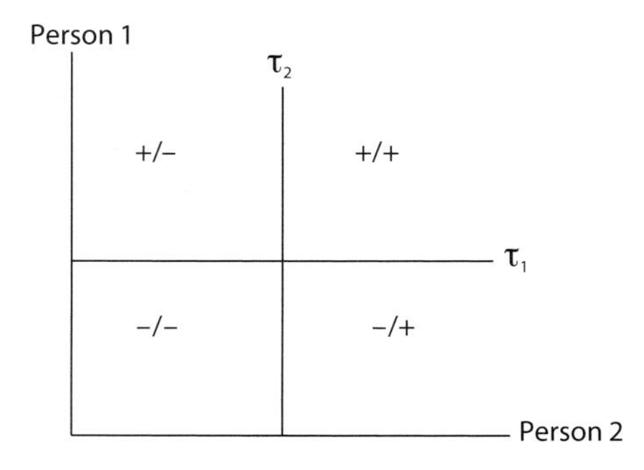

FIGURE 13. Two Thresholds Implies Four Regions

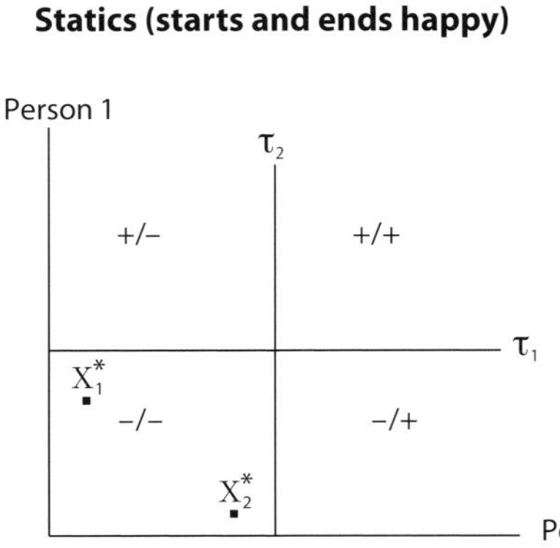

Statics (starts and ends happy)

Person 1

τ_2

+/– +/+

τ_1

X_1^* –/– –/+

X_2^*

Person 2

Figure 14. Equilibria Benign

Dynamics! (Gets bad en route)

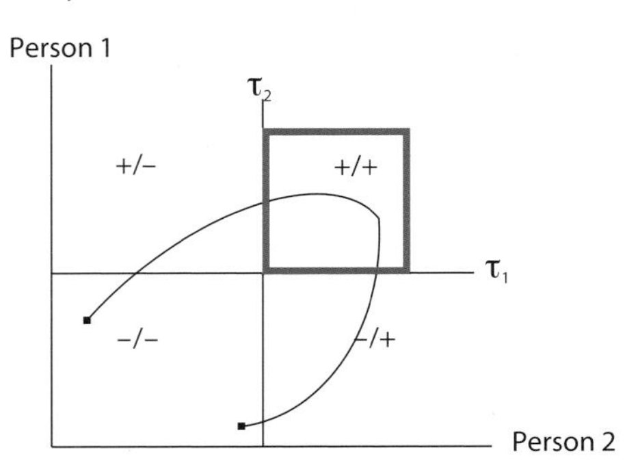

Figure 15. Out of Equilibrium Dynamics Not Benign

into four regions. In the +/+ region both agents are above threshold, while in the −/+ region only person 2 is, and so on.

If we imagine the action to be internment (or worse) of a feared group, for instance, then comparative statics (comparison only of equilibria without dynamics) is not enough. We can easily imagine a process beginning and ending in the benign state (−/−), where neither supports internment, with the points X_1^* and X_2^* denoting the agents' respective views in equilibrium as illustrated in Figure 14.

But this does not preclude them passing through a brutal patch en route, as shown in Figure 15.

So, we care about dispositional *dynamics*, not just equilibria (if such even exist).[88]

Coupled Dispositional Trajectories

Returning, then, to our two individuals in an exogenous environment of trials (e.g., attacks by violent jihadists[89]), the coupled model works as follows. First, each agent "solves" (speaking figuratively—see the next section) the Rescorla-Wagner model for his purely affective trajectory, conditioning on the stream of trials (one per time unit). Second, at each time, he computes (what he believes to be) the conditional probability of violent jihadist

[88]The distinction between the mere existence of equilibria and their dynamic attainment in social systems is fundamental to the generative epistemology. For extended discussions see J. M. Epstein, 2006.

[89]The term *jihad* admits a number of meanings, one of which is Holy War and another of which is internal struggle to obey the Koran.

given Muslim. The sum of these, as before, is his solo disposition, $V + P$. Third, he applies a weight to the solo disposition $(V + P)$ of the other person and adds this to his own. The second person does likewise, but with his own parameters and his own weighting of person 1's solo disposition.

These weights define *a dispositional network*. Each then subtracts his threshold from that sum, acting (attacking a random Muslim) if the result is positive and not acting otherwise. Obviously, one can imagine the subjects as minority Americans, the conditioning trials being attacks by white supremacists, with persecuted agents estimating the conditional probability of white supremacist given white—or people estimating the probability of dangerous vaccine given vaccine, or corrupt executive given executive, or disgusting uni[90] given uni, or transporting kiss given kiss, and so forth.

Mathematically, this is an unusual combination of differential equations and algebra, in that a set of differential equations is first solved and the solutions are then superposed in a weighted sum.[91]

Solve vs. Conform To

I am emphatically *not* suggesting that any of these calculations are consciously executed by the individual, that the individual is assumed to know differential equations, or any such thing. I am articulating an algorithm to which the agents are hypothesized to conform. The eagle *conforms to* the equations of aerodynamics but is obviously not *solving* them. Likewise, people conform to grammatical rules they certainly cannot articulate. Indeed, they are blissfully unaware that they are following them at all.[92] The mere fact that typical people cannot *solve* the optimization equations of economics does not per se dismiss them as predictors of behavior. The two-person setup is summarized in Figure 16.

With zero weights, we recover the decoupled agents, resolving their separate internal rivalries between passion V and reason P.

With this simplest of all networks in hand, we can construct innumerable cases of the sort imagined earlier—in which both agents begin in the benign $(-/-)$ state, passing through the split $(-/+)$ state, and arriving in the active $(+/+)$ state. The parameters given are quite arbitrary and were selected merely as an existence proof for such interdependent dispositional trajectories. Figure 17 offers one example. Here, we graph disposition net of threshold, $D^{net} = D^{tot} - \tau$, so action is taken if and only if curves

[90] Urchin sushi.

[91] Earlier model drafts tried putting the threshold term in the differential equations, for example. But if one simply adds τ to the right-hand side, the solution, $v(t)$, becomes unbounded.

[92] As powerful demonstrations of how difficult it is to characterize these rules formally, see S. D. Epstein (2000) and S. D. Epstein and Seely (2002).

$$\frac{dv_1}{dt} = \alpha_1 \beta_1 (\lambda - v_1) \qquad [21]$$
$$\frac{dv_2}{dt} = \alpha_2 \beta_2 (\lambda - v_2)$$
$$v_i(0) = v_0$$

Weights define dispositional network. Extract v-functions and compute net dispositions:

$$D_1^{\text{net}}(t) = v_1(t) + P_1 + \omega_{21}(v_2(t) + P_2) - \tau_1 \qquad [22]$$
$$D_2^{\text{net}}(t) = v_2(t) + P_2 + \omega_{12}(v_1(t) + P_1) - \tau_2$$

FIGURE 16. Two-Agent Coupled Model

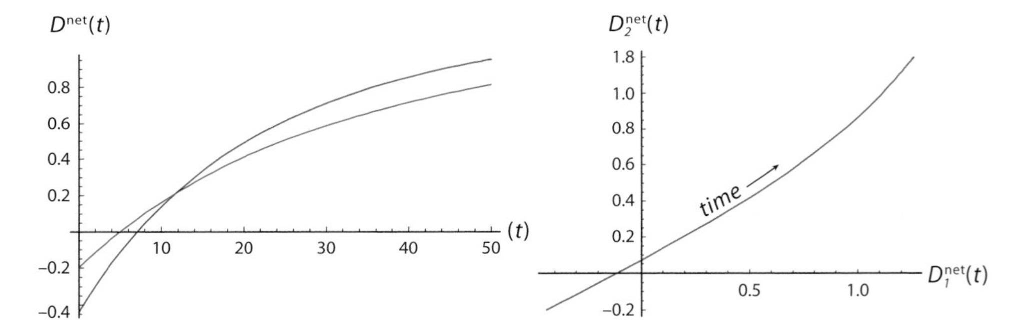

FIGURE 17. From Both Negative to Both Positive

break the x-axis, as per action rule [4]. *Mathematica* Code is provided in Appendix II and on the book's Princeton University Press Website.

Both net dispositions begin negative in the left graph and, so, in the $-/-$ quadrant of the right graph. Then, on the left, while Blue is still negative, Red crosses his threshold at $t = 5$. This continues through the period $5 < t < 8$, which is in the $+/-$ quadrant on the right. Then, finally, at roughly $t = 8$, Blue goes positive and we arrive in the $+/+$ quadrant.

We embed this in a much richer story in Figure 18, which subsumes, in the simplest way, the core (e.g., lynching) example given in the Introduction. Richer renditions of that story will be presented in the three-agent mathematical versions to come and then spatiotemporally in the full agent-based model of Part II.

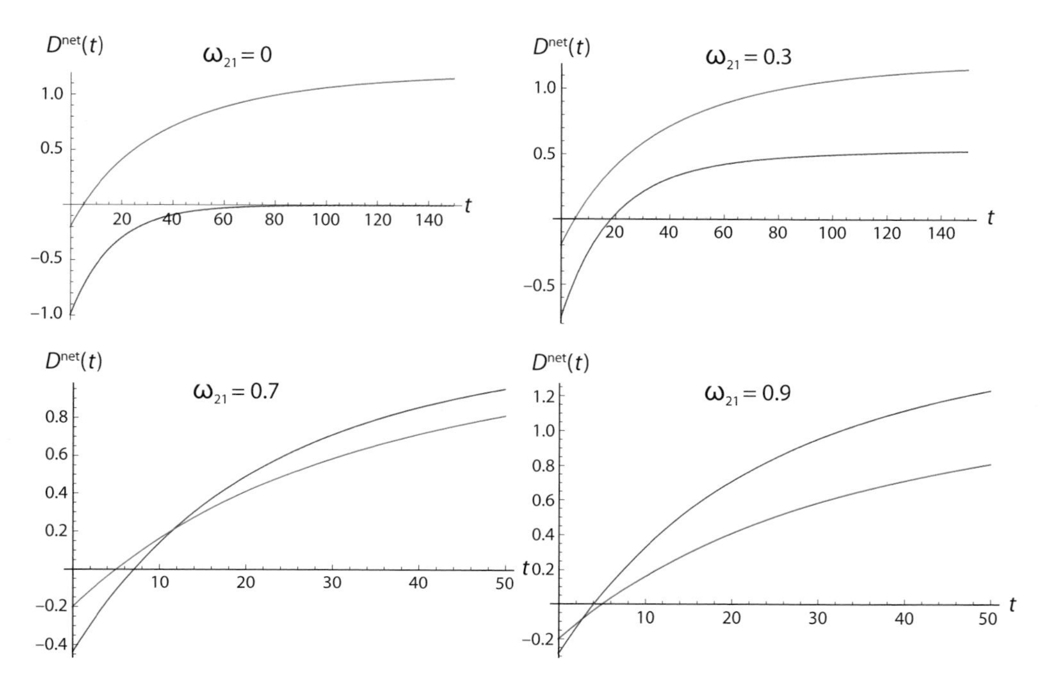

FIGURE 18. Dispositional Contagion

Simple Version of the Core Target

Referring to Figure 18 we see that, of his own volition, Agent 1 (Blue) would never act. This is shown in the upper-left frame, where ω_{21} (the weight of 2 on 1) is zero. Agent 1's initial affect and his immediate probability estimates are both zero. And, for these parameters, the Blue net disposition curve is never positive.[93]

If we hike ω_{21} to 0.3, Agent 1's total disposition does exceed threshold ($D^{net} > 0$), but he goes second and never exceeds Agent 2. The third frame ($\omega_{21} = 0.7$) was discussed earlier. Here, Agent 1 surpasses Agent 2 but still does not *initiate* the action. Finally, with $\omega_{21} = 0.9$ Agent 1 acts *first* and remains the most virulently committed throughout.[94]

[93]This holds because the least upper bound is $\lambda - \tau$ and, since here they are both 1, the asymptotic value is zero. Note also that, if we fix both Agent 1's affect (V) and probability (P) to zero, then his disposition trajectory is identical to Agent 2's. So, while his initial affect is zero, we do give Agent 1 a small learning rate ($\alpha\beta$) so he has a distinguishable disposition trajectory. But this still does not exceed threshold. Of his own volition, he would never act. However, with dispositional contagion, he is the first to act, which is our target.

[94]These cases unfold at quite different rates. Hence, for visual purposes, the t-axes differ among them.

Now, in panel 4, where Agent 1 acts first, his observable behavior certainly suggests "leadership." But, as we see, he is simply the most susceptible to dispositional contagion.[95] We will return to this distinction and the general discussion of leadership shortly.

Notice the role of the Rescorla-Wagner model in this two-agent story. If we cancel it (by setting Agent 1's learning rate to zero) and also keep Agent 1's probability (P) pegged at zero as before, then Agent 1's net disposition reduces to

$$D_1^{\text{net}} = \omega_{21}(V_2 + P_2) - \tau_1. \qquad [23]$$

His highest disposition trajectory is obviously attained when $\omega_{21} = 1$. But, retaining our usual assumption of equal thresholds, this makes Agent 1 identical to Agent 2. This is still noteworthy in that alone, Agent 1 would sit at $-\tau_1$, whereas with $\omega_{21} = 1$, he goose-steps across the threshold arm in arm with Agent 2. This is far from trivial. But with zero learning, he cannot go first—in the *two*-person case! As we shall see, in the three-person case, this is possible despite the fact that solo disposition net of threshold is negative. This is an important difference between the two-person and larger population cases.[96]

The Full Trajectory

Returning now to the mathematical model development, all of this has had to do with the acquisition phase of Figure 15—the transition from −/− to the red zone of +/+. But the full trajectory returns to −/−. By what mechanism? Dispositional extinction, as illustrated in Figure 19, which picks up where Frame 4 of Figure 18 leaves off ($t = 50$). To be precise, this is the net disposition trajectory under affective extinction, as treated in the Rescorla-Wager model.[97] Of course, the agents are coupled through the weights. So, the extinction process, like the acquisition phase, would also exhibit contagion effects. But this example offers one realization of the troubling full trajectory of Figure 15. Notice that, as discussed earlier, the Blue agent's dispositional extinction is steeper in Figure 19 than that of the Red agent. Both dispositions converge to asymptotic values because, as in

[95] This is to be distinguished from the case where the first to act simply has a lower threshold than everyone else. This may occur, but it is not the case here.

[96] One could explore what Agent 1 learning trajectory, in a direction counter to Agent 2's, is necessary to counter the latter's influence, permitting 1 to stay negative.

[97] The affective extinction component is handled in the Rescorla-Wagner model by setting λ to zero. This pulls down the disposition, as shown. But the disposition still includes the agents' P-values and thresholds, τ . Thus, a subtlety of the extinction calculation is that the extinction rate applies to the purely affective components of the disposition, which in Figure 18, Frame 4, are roughly .86 and .60 when extinction begins. See Appendix II for full *Mathematica* programs.

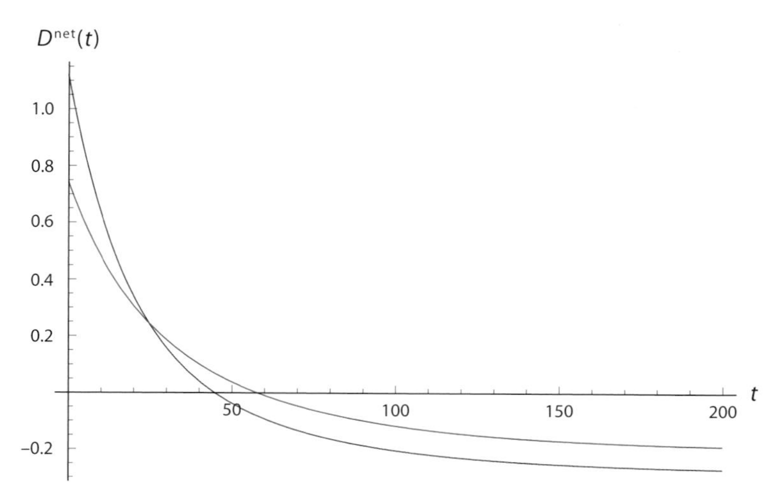

FIGURE 19. Dispositional Extinction to Benign State

equation [16], affects extinguish exponentially to zero, and all other net disposition terms are constants.

Examples of Fear Contagion

Having shown that the basic model can generate them, is there actually evidence of interdependent affective trajectories, that a person can acquire fear through exposure to a conspecific's expressions thereof, without direct experience of an aversive stimulus?

There is evidence of many sorts. Before reviewing it, we recall the earlier point that this capacity to acquire fear without direct conditioning is advantageous evolutionarily. "In particular, humans and other primates have developed social means of acquiring fears, which permit learning about the potential aversive properties of stimuli in the environment without necessarily having to experience the aversive event directly" (Ledoux and Phelps, 2008, p. 67).

Historical

Anecdotal evidence of emotional contagion is, of course, abundant. Gustave Le Bon (1895; 2001 ed.) wrote, "Sentiments, emotions, and ideas possess in crowds a contagious power as intense as that of microbes."[98] A panoply of contagious panics, phobias, and general hysterias is catalogued in Mackay's

[98] Quoted in Hatfield, Cacioppo, and Rapson (1994, p. 105).

compendious *Extraordinary Popular Delusions and the Madness of Crowds* (1841). Many of these—but most particularly the witch manias—are obviously fear based[99] and eventuated in gruesome violence. A prominent example is the Salem Witch mania, although the contagious fear and mutual suspicion of McCarthy's witch hunts were no less virulent.

The fear-driven cat massacre of 1884 is a colorful example. This was stimulated by an outbreak of plague, presumed to be caused not by rats, but by cats. The ensuing cat massacre (in exterminating the rats' main predator) only exacerbated the problem. This and many other historical examples are discussed in Hatfield, Cacioppo, and Rapson's *Emotional Contagion* (1998).

A wealth of historical emotional contagion evidence from the field of public health is reviewed in J. M. Epstein et al. (2008), which develops mathematical and agent-based models coupling the contagion dynamics of fear and disease. Specifically, the model posits two epidemics: one of disease proper and one of *fear of* the disease. Fear stimulates changes in peoples' contact behaviors (e.g., flight or self-isolation), and these behavioral adaptations feed back to alter the disease trajectory. The coupled model offers a new, behavioral, mechanism to explain the multiple epidemic waves observed in the 1918 pandemic flu—a dynamic of long-standing interest to epidemiologists.

The R_0 of Fear

A fundamental parameter in epidemiology is the so-called basic reproductive rate of a disease. Termed the R_0—or R-naught—it is defined as the number of secondary infections resulting when a single infectious individual is placed in a population of susceptible individuals. In J. M. Epstein et al. (2008) we also develop a theoretical expression for the R_0 of fear and give the conditions under which it exceeds the R_0 of the underlying disease. Indeed, unlike prevalence-dependent models (Kremer, 1996; Philipson, 2000), fear can spread in the absence of disease.

An arresting contemporary health example of this is the case of Surat, India. "In 1994, hundreds of thousands of people fled the Indian city of Surat to escape pneumonic plague, although by World Health Organization criteria no cases were confirmed" (see J. M. Epstein, 2009). Fear-based refusal of effective vaccines is a related problem of widespread concern; it is central to the resurgence of polio, for example. In addition to violence and public health, the cascading financial crisis of 2008–2009 was clearly accelerated by a contagion of fear concerning the solvency of major financial institutions. Earlier examples of financial panics are discussed in Charles

[99] I say "obviously" because I presume there are, in fact, no witches. But that is probably a minority view.

Kindleberger's classic *Manias, Panics, and Crashes* (1978). Even olfactory stimuli are found: a sudden-onset epidemic of hysteria among schoolchildren, stimulated by an odor (!) is presented in G. W. Small et al. (1994). For further fascinating examples of what he terms the "social transmission of psychopathology," see William Eaton's *The Sociology of Mental Disorders* (2001). Recent research indicates that chemo-signals can also communicate human emotions, including fear and disgust (de Groot et al., 2012).

Adam Smith certainly believed in emotional contagion. In *The Theory of Moral Sentiments* (1759; 1982 ed.), he wrote, "the passions, upon some occasions, may seem to be transfused from one man to another, instantaneously and antecedent to any knowledge of what excited them in the first place."

But, beyond historical analyses and innumerable anecdotes, there is also rigorous neuroscience to identify the underlying mechanisms.

Mechanisms of Fear Contagion

Jamesian Mechanisms—Facial and Postural Mimicry

Much of this work stems from the amazingly prescient theory of William James (1884), which centered on the view that unconscious mimicry of facial expressions stimulates emotional convergence: we *first* unconsciously mimic the other's facial expression; our facial expression then triggers our emotion.[100] On the latter point, the classic experiment is Strack, Martin, and Stepper (1988; see also Adelmann and Zajonc, 1989). A nice discussion is given by Kahneman (2011). The encoding of facial expressions into emotions is apparently innate and is observed even in humans blind from birth and thus unable to acquire this mapping by observation (Haviland-Jones and Wilson, 2008, p. 238). Indirect suggestive evidence to this effect is that motor disorders, such as Parkinson's and Huntington's disease, are associated with emotional deficits, perhaps because the actual emotional stimulus, one's adoption of a facial expression, is blocked (de Gelder et al., 2004). The substantial literature on fear contagion through facial expression has been extended by de Gelder et al. to full body postures, "similar to what has so far been argued for automatic recognition of fear in facial expressions" (p. 16703). In fact, in very interesting recent experiments, Aviezer, Trope, and Todorov (2012) found that body cues can dominate facial expressions in our discrimination of others' emotions. For a compact history of James's

[100] In terms of the model, facial mimicry is not the imitation of a consequential binary behavior or action (e.g., join the lynch mob). It can contribute to the construction of one's disposition. But it is the overall disposition's relation to threshold that triggers *action*, as the latter is defined in this model.

ideas and their successors, see LeDoux (2009). A central work in this tradition is Hatfield, Cacioppo, and Rapson (1994).

Laboratory Neuroscience

Current neuroimaging studies shed further light on the acquisition of emotions such as fear through observation. Olsson, Nearing, and Phelps (2007, p. 1) report that "classical fear conditioning requires first-hand experience with an aversive event, which may not be how most fears are acquired in humans." They write that "fear acquired indirectly through social observation, with no personal experience of the aversive event, engages similar neural mechanisms as fear conditioning." Their study suggests that "indirectly attained fears may be as powerful as fears originating from direct experiences." As earlier noted, the capacity to acquire fear by observation is highly adaptive, of course. Fear acquisition through "social observation and verbal communication" are "more efficient and associated with fewer risks than learning through direct aversive experience."

The amygdala is centrally implicated in both direct and observational fear acquisition (LeDoux, 1996). In monkeys, observational fear learning has been shown to be similar to direct fear conditioning. Olsson, Nearing, and Phelps (2007) report, "In particular, work on observational fear-learning in monkeys has shown that the relationships between the magnitude of a learning model's expressed distress, the observer's immediate response to the model's distress and the resulting fear-learning in the observer are similar to those existing between an US, UR and a CR in classical fear conditioning paradigms (Cook and Mineka, 1990; Mineka and Cook, 1993)." They continue, "A recent study directly comparing human fear-learning through conditioning, social observation and verbal instruction supports the same conclusion (Olsson and Phelps, 2004)."

The following experiment is very impressive and is described in full detail in Olsson, Nearing, and Phelps (2007). Summarizing, a set of individuals were directly fear-conditioned by repeated pairings of a blue square (the CS) and an electric shock (the US). The CR was measured both by skin conductance response (CSR) and by functional magnetic resonance imaging (fMRI). This direct fear-conditioning exercise was filmed. Then, the true subjects were instructed to *watch* the movie. After having done so, they received the same CS (the blue square) and, remarkably, their CRs were comparable to those of the directly conditioned individuals.

"Subjects in our study showed a robust fear response following observation, corroborating previous reports of comparable behavioral (Mineka et al., 1984; Mineka and Cook, 1993) and psychophysiological (Olsson and Phelps, 2004) expressions of fear following observational learning and fear conditioning" (Olsson, Nearing, and Phelps, 2007, p. 8).

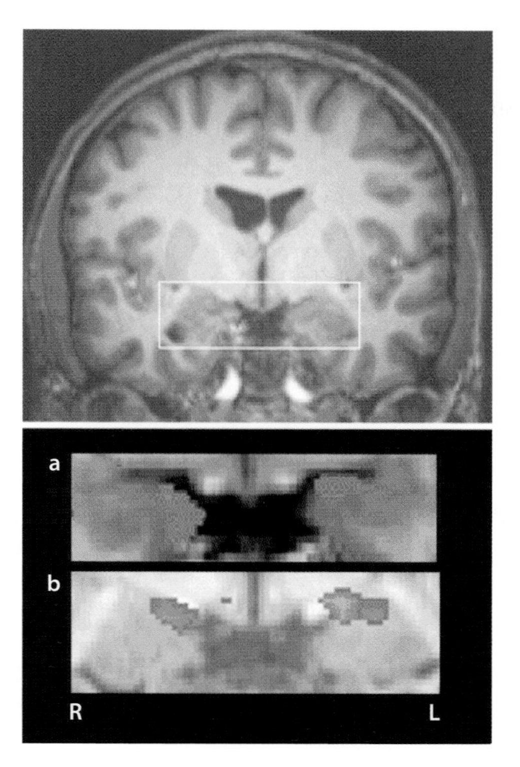

Figure 20. Observational Fear Conditioning. Source: Reprinted by permission from Macmillan Publishers Ltd: *Nature Neuroscience* (Olsson and Phelps 2007, p. 289), copyright 2007

Neuroanatomically, they add: "Importantly, our imaging data provides the first evidence that the amygdala, which is known to be critical to the acquisition and expression of conditioned fear (Phelps and LeDoux, 2005), is similarly recruited during the acquisition and expression of fear acquired indirectly through social observation" (Olsson, Nearing, and Phelps, 2007, p. 6).

Specifically, Panel a of Figure 20 shows the excitation pattern in the earlier Figure 5 for a subject after direct fear conditioning. Panel b shows the excitation pattern for a subject *observing* that same process of fear conditioning. Having watched that process, in other words, this is the observer's response to the blue square (the CS). Again, detailed analysis of the comparison is given in LeDoux and Phelps (2008).

"Our finding that the formation and the expressions of fear through social observation relies on neural circuits that are similarly involved in fear conditioning is in accordance with the description of this form of learning as an evolutionarily old system for the transmission of emotionally relevant information, as documented in a wide range of species" (Olsson, Nearing, and Phelps, 2007, p. 10).

In sum, the Olsson, Nearing, and Phelps (2007, p. 10) results "show that fears learned by observing others engage the same neural mechanisms as fear acquired through direct experience, suggesting that social and nonsocial means of fear learning may be equally effective and powerful."

So, yes, there is evidence that an individual's fear may be "contracted" by another, without direct exposure to the aversive conditioning stimulus. Recent work on mirror neurons offers yet further reinforcement for emotional contagion, as do lesion studies (explored in Part II), "knock-out" animal models, and a wide variety of other approaches.

Mirror Neurons

Of such selective advantage is this capacity to acquire fear indirectly that we *may* even have evolved specialized neurons—so-called *mirror neurons*—for this purpose. Rizzolati and Fabbri-Desto contend that "Mirror neurons are a distinct class of motor neurons that discharge both when individuals perform a specific motor act *and* when they observe the same act done by another individual" (Rizzolati and Fabbri-Desto, 2009, p. 337). See, for example, Rizzolatti and Craighero (2004), Iacoboni (2009), and Caramazza (2011).[101]

Experimental work by Gazzola, Van der Worp et al. (2007) suggests that actual imitative capacity is not necessary for the recruitment of relevant areas. Here, "Two aplasic individuals, born without arms or hands, were scanned while they observed hand actions. The results showed activations in the [usual] parieto-frontal circuit of the aplasic individuals while watching hand actions" (p. 344).

Contagion Beyond Immediate Vision

Emotional contagion can certainly occur outside of visual range. Obviously, blind people can acquire others' fears through vocal cues, without literally seeing them. But emotional contagion can occur at any distance, by cell phone, field radio, text message, e-mail, tweet, and so forth.[102]

[101] As a matter of sheer speculation, I wonder if mirror neurons explain why we laugh at pratfall humor. We know, through mirror neurons, how these incidents feel, and are laughing with relief as they befall someone else!

[102] In monkeys, sounds expressing fear stimulate fright responses (freezing, flight). And in humans, communication (and dispositional contagion) can be vocal. On the vocal communication of emotion in humans, see Pittam and Scherer (1993, p. 340). We discussed the auditory route to the amygdala previously. Communications of emotion by text is another route, as reported in Funayama et al. (2001).

There is no question that, beyond cell phones, social media of all sorts played a major role in spreading the Arab Spring (discussed in Part III) and has been a central medium in the spread of widespread reactions to isolated events. For example, in 2012 a single independent YouTube video defamatory of Islam precipitated mass reactions across the Middle East, Southwest Asia, Indonesia, and Europe.[103] It is implausible that the majority of participants ever saw the video (were ever exposed to the primary stimulus), indicating that the disposition to protest was transmitted at long range by other means.

Heterogeneity

Finally, people exhibit heterogeneous susceptibility to emotional contagion. Hatfield, Cacioppo, and Rapson (1994) and Doherty (1998) develop an emotional contagion scale and demonstrate substantial variation across samples. The heterogeneous weights used in the mathematical and agent-based models here register this heterogeneity, which will be shown to play a central role in our understanding of so-called leadership and the evolution of collective behavior.

In sum, the phenomenon of emotional contagion is powerfully documented, and a diversity of neural mechanisms has been identified. Thus, *while we are not modeling the brain proper, the brain science licenses the modeling*. Affective contagion dynamics are crudely represented in the equations and are represented even more vividly in the agent-based model of Part II.

In the model, the probabilistic, nonemotional component of our behavior can also display strong contagion, or conformity, effects. Is this, too, supported by the neuroscience?

Conformist Empirical Estimates

The remarkable fact is that even a person's empirical judgments are affected by those of other people. The most arresting example is the famous Asch (1956) experiment. Here, the mathematical judgment is much more primitive: Are two line segments of equal length? Asch's hand-drawn original test figure is reproduced in Figure 21.

Obviously, in Asch's experiment the lines did not carry labels betraying their true lengths (8 inches, and so on). Without them, the subject was shown the lines in Figure 21 and was asked to simply identify that line on the right that is the same length as the one on the left. When questioned alone (the control group), only 2 in 37 gave an incorrect answer. But in the

[103]The bombing of the U.S. Embassy in Benghazi was distinct from these mass spontaneous demonstrations.

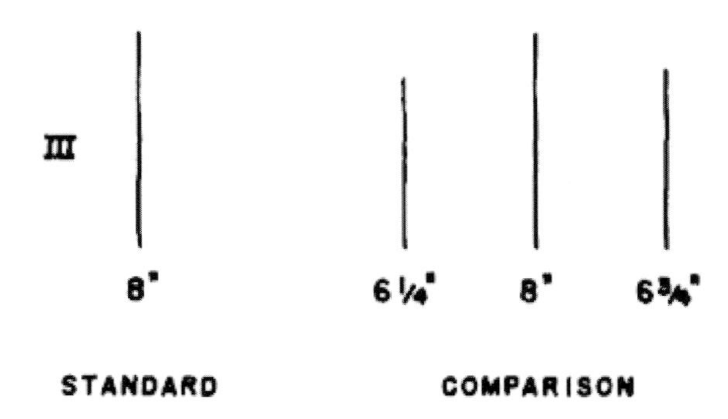

Figure 21. Critical Comparisons. Source: Asch (1956)

presence of majorities unanimously giving erroneous answers,[104] the subjects' judgments were strikingly different, as Asch (1956) reported:

> The estimates of the control group were virtually free of error. Thirty-five of 37 subjects made errorless estimates; of the remaining two subjects one showed one error, the other two errors. The proportion of errors was less than 1 per cent of the total number of critical estimates.
>
> In contrast, the critical subjects showed a marked movement toward the majority. Errors increased strikingly, their frequency among individual subjects ranging from 0 to 12, or up to the maximum the conditions permitted. *Only one-fourth of the subjects in the three experimental groups showed errorless performances, while in the control group 95 per cent were free of error.* [Emphasis added.]

One might imagine that the subjects knew the right answers, and were just voicing the incorrect majority view to "go along." This was certainly not the case uniformly. Having been told of their mistake in postexperimental interviews, many of the subjects concluded that their eyes or visual processing must have failed them, or that their head might have been tilted to produce an erroneous answer. But they did *not* deny that they believed what they had reported (Asch, 1956).

So, in fact, there are numerous sources of inaccuracy in the agent's probability estimate: there is (a) sample selection bias, which may be damped or amplified by (b) memory, which may be damped or amplified by the (c) weighted estimates of others.

[104] Asch varied the size of these confederate majorities but found that a majority of 3 was sufficient to produce a powerful "majority effect," as he called this phenomenon.

Asch, of course, did not have the benefit of contemporary neural imaging technology and hence was not able to suggest neurobiological correlates of conformity.[105] Recent studies—in addition to demonstrating arresting conformity effects—are beginning to do so.

Neural Basis for Conformity Effects

Berns et al. (2005) use a three-dimensional figure rather than Asch's lines to clearly show conformity effects and the regions of interest in their connection.[106] In a fascinating section entitled, "The Pain of Independence," (p. 252) the authors report, "The amygdala activation in our experiment was perhaps the clearest marker of the emotional load associated with standing up for one's beliefs."

The term *pain* is proving to be entirely justified. Recent work by Kross et al. (2011, p. 6270) shows that " . . . when rejection is powerfully elicited . . . areas that support the sensory components of physical pain (secondary somatosensory cortex; dorsal posterior insula) become active." The authors ". . . demonstrate the overlap between social rejection and physical pain in these areas by comparing both conditions in the same individuals using fMRI." They find that "activation of these regions was highly diagnostic of physical pain, with positive predictive value of 88%." As they write, "These results give new meaning to the idea that rejection 'hurts' . . . rejection and physical pain are similar not only in that they are distressing—they share a common somatosensory representation as well."

Their "whole brain" and "region of interest (ROI)" analyses are shown in Figure 22. Particularly arresting in both cases is the overlap between social rejection and physical pain in S2 (the secondary somatosensory cortex).

This is highly original work. Kross et al. (2011, p. 6272) appropriately report that ". . . intense social rejection activated somatosensory regions that are strongly associated with physical pain, which are virtually never associated with emotion as typically studied." Earlier, we discussed the evidence for observational fear conditioning (Olsson, Nearing, and Phelps, 2007). Kross et al. (p. 6274) raise the parallel question, "whether observing someone else experiencing intense social (rather than physical) pain (i.e., a parent witnessing their child's rejection) also recruits sensory-pain processing regions." This would help explain why *public* displays of social rejection—such as shunning—and, specifically, *public* punishment displays (e.g., the town-square pillory) are so effective in enforcing norms.

[105] On priming and conformity, see Epley and Gilovich (1999).

[106] These are the amygdale [sic] and the right caudate nucleus. For quite different work on conformity, not involving geometric comparisons, see Spitzer et al. (2007, pp. 185–196).

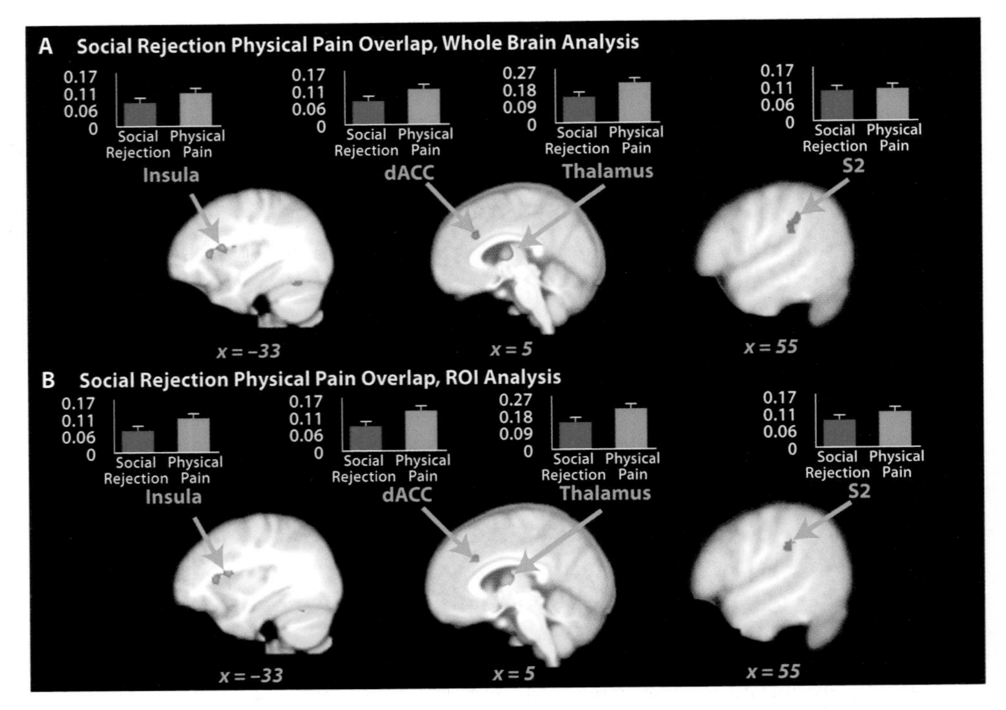

FIGURE 22. Social Rejection and Physical Pain. Neural overlap between social rejection and physical pain. (*A*) A whole-brain conjunction analysis revealed that regions typically involved in both the affective [AI (−30, 11, 14); dACC (9, 26, 24)] and sensory [thalamus (6, −4, 7); S2 (62, −28, 36)] components of physical pain were also involved in response to social rejection (ex-partner > friend) and physical pain (hot > warm). (*B*) An ROI analysis performed on physical pain regions revealed overlap between social rejection and physical pain in regions similar to those identified by the whole-brain analysis [AI (−33, 11, 14); dACC (6, 26, 24); thalamus (6, −4, 7); S2 (59, −26, 24)]. Bar graphs demonstrate the β-values for social rejection (ex-partner > friend) and physical pain (hot > warm) extracted from each cluster. Error bars represent one SE. None of the β-values associated with social rejection differed significantly from the β-values associated with physical pain (all two-tailed paired sample t statistics < 1.75, all P values > 0.09). Source: Kross et al. (2011, p. 6271); reprinted courtesy of PNAS

Evolutionarily, one might hypothesize that conformity is required for organized group defense. Groups incapable of powerful conformity are selected out in the intergroup competition. Hence, groups whose members are equipped with powerful wiring in this conformist direction—as where social rejection physically hurts—will enjoy selective advantage and are the ones that will be observed, as they are in Figure 22. A different, individual-level, evolutionary argument would be that social ostracism undermines mating opportunities. The species thus evolves mechanisms making it physically painful to engage in actions that threaten social acceptance and mating opportunities—literal self-punishment and deterrence are at work inside the individual.

As suggested earlier, however, the cost of this bias toward conformity is that it suppresses individual innovation and (painful) independence. Understanding this evolved source of our conformist impulses may equip us to overcome them, expressing the dissent so essential to democracy.

Thus, not only is there a vast psychology literature establishing conformity effects, but there is emerging a fascinating literature explaining them at a deeper neural level. Emboldened by all of this research, then, let us proceed to further elaborate the mathematical model assuming conformity effects, en route to the full spatial agent-based treatment of Part II. Many of the preceding topics will arise again in developing the agent-based model to follow. Hopefully, the foregoing historical, psychological, and neuroscience discussions are sufficient to justify the continuing model exposition.

Generalizing Rescorla-Wagner

This is a reasonable place to generalize the Rescorla-Wagner model slightly to accommodate a wider variety of learning trajectories. The generalized one-person model is shown in Figure 23.

The original model is the special case of $\delta = 0$. The case, $\delta = 1$ yields classic s-curve learning. Both these cases are analytically solvable. For δ strictly between 0 and 1, numerical methods are needed. I will use *Mathematica's* nonlinear differential equation solver NDSolve, but a standard numerical

$$\frac{dv}{dt} = \alpha\beta v^{\delta}(\lambda - v) \qquad [24]$$

$\delta = 0$ Rescorla-Wagner (analytically solvable)

$\delta = 1$ Classical Logistic S-curve (analytically solvable)

$\delta \in (0,1)$ Other S-curves (not analytically solvable)

FIGURE 23. Generalizing Rescorla-Wagner

approach like fourth-order Runge-Kutta will give indistinguishable results. All *Mathematica* Code is given in Appendix II and on the book's Princeton University Press Website.

Nontrivial Majorities: The Three-Agent Model

We have discussed the two-agent model. But, for nontrivial forms of group decision, such as voting systems or juries, we require three or more individuals. This allows majority rule, which we model next. Extending the earlier two-person model directly, we introduce a third agent, as depicted in Figure 24. All agents are now equipped with the generalized model.

To see how this works, let us walk through the algorithm from the perspective of Agent 1, understanding that I am not claiming that any of this is knowingly executed by any human. The claim, as always, is that *Agent_Zero* is theoretically fruitful, not that it is descriptive.

With this in mind, let us colloquially say that person 1 first "solves" her generalized Rescorla-Wagner differential equation for $v(t)$, where t indexes an exogenous stream of direct conditioning trials, which may be null. She then computes her probability estimate P_1 (in the subsequent spatial agent model, this will be the result of local sampling). If she placed no weight on the affect or probability estimates of others, she would simply subtract her threshold from this sum and act accordingly. But, in general, she "catches" the dispositions of others. The six weights indicated in the diagram register the fraction of the others' (solo) dispositions contracted by each agent—through facial expressions, vocal cues, text messages, or other neurosocial mechanisms, as discussed earlier. This overall sum—the total disposition—is what each agent "compares" to his or her threshold.

$$\frac{dv_1}{dt} = \alpha_1 \beta_1 v_1^{\delta_1} (\lambda - v_1) \qquad [25]$$

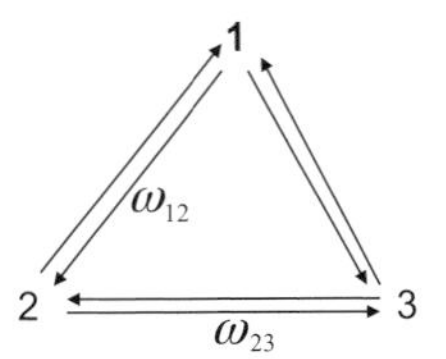

$$\frac{dv_2}{dt} = \alpha_2 \beta_2 v_2^{\delta_2} (\lambda - v_2)$$

$$\frac{dv_3}{dt} = \alpha_3 \beta_3 v_3^{\delta_3} (\lambda - v_3)$$

$$v_i(0) = v_0$$

Weights define dispositional network. Extract v-functions and compute net dispositions:

$$D_1^{\text{net}}(t) = v_1(t) + P_1 + \omega_{21}(v_2(t) + P_2) + \omega_{31}(v_3(t) + P_3) - \tau_1 \qquad [26]$$

$$D_2^{\text{net}}(t) = v_2(t) + P_2 + \omega_{12}(v_1(t) + P_1) + \omega_{32}(v_3(t) + P_3) - \tau_2$$

$$D_3^{\text{net}}(t) = v_3(t) + P_3 + \omega_{13}(v_1(t) + P_1) + \omega_{23}(v_2(t) + P_2) - \tau_3$$

Figure 24. Generalized Three-Agent Model

The Central Case

I will now show two mathematical versions of our central case: the person who leads the lynch mob despite (a) harboring no aversion to black people (v is always zero), (b) having no evidence of black wrongdoing (P is always zero), (c) the same action threshold as everyone else, and (d) no orders.

Case 1: Homogeneous Classical Rescorla-Wagner Learners (All δ's Zero)

First, I have made Agents 1 (Blue) and 2 (Red) literally identical to produce the simplest version. So their (rising) curves coincide (appearing as a single purple curve). If Agent 3 (Green) is completely invulnerable to dispositional contagion ($\omega_{13} = \omega_{23} = 0$), then his disposition simply sits at $-\tau$, as shown in the left panel of Figure 25. He would never act. However, if he is maximally susceptible to dispositional contagion ($\omega_{13} = \omega_{23} = 1$), we generate the trajectory on the right, in which he goes first.

This is the simplest version of the central result. However, we have not exploited the generalized version of the model, which allows heterogeneity in learning modes (e.g., *S*-curve learning). This interesting heterogeneity comes from the δ's.

Case 2: Heterogeneous Nonclassical Learners

Introducing $\delta > 0$ enriches the learning trajectories substantially, as we shall see.[107] Again, we show two runs (Figure 26).

In the left panel, our protagonist (Green Agent 3), left to his own devices (i.e., with $\omega_{13} = \omega_{23} = 0$) would never act—net disposition is negative throughout, and his empirical probability estimate is zero throughout. However, with positive weight on the other agents' (solo) dispositions, Agent 3 again initiates the lynch mob, as shown in the right panel. Agents 1 and 2 are *S*-curve learners, with δ's of 1.0 and 0.5, respectively (see Appendix II code). This produces much more interweaving of trajectories, including a period between 100 and 150, in which Agent 3 relinquishes, but then regains, the lead. Again, the essential point is that Agent 3 begins with the strongest *negative* disposition, but, through dispositional contagion, acts first, and after some jockeying with Agent 1, remains the most

[107] However, this does impose a subtle constraint on the initial values of v. Unless one wishes to suppress all learning, these can be imperceptibly small—but cannot be literally zero. If $\delta > 0$, then $v_0 = 0$ suppresses all learning. This is easiest to appreciate in the difference equation version. The first v-update would be $v_1 = v_0 + \alpha\beta v_0^{\delta}(\lambda - v_0)$. If $\delta > 0$, there is no update because 0^{δ} is 0. With $\delta = 0$, the update is $\alpha\beta\lambda$, because v_0^{δ} is then 0^0, which is 1. In the runs of Figure 26, I use 0.0001, which is effectively 0 in this context. For comparability, I use the same value in Figure 25.

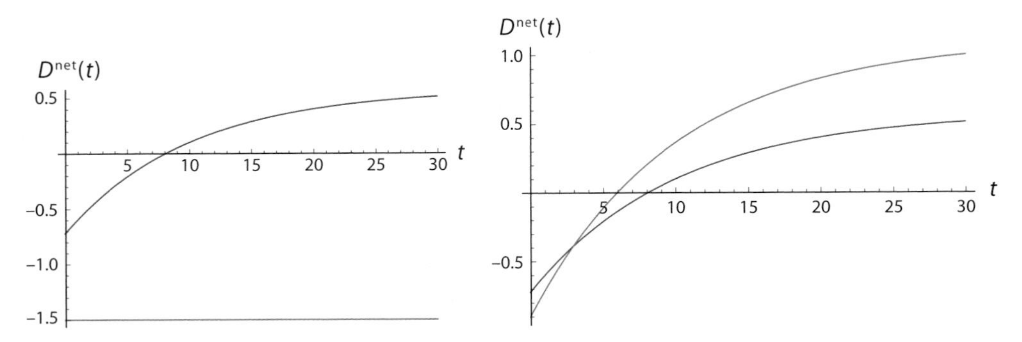

FIGURE 25. Homogeneous Classical Learners

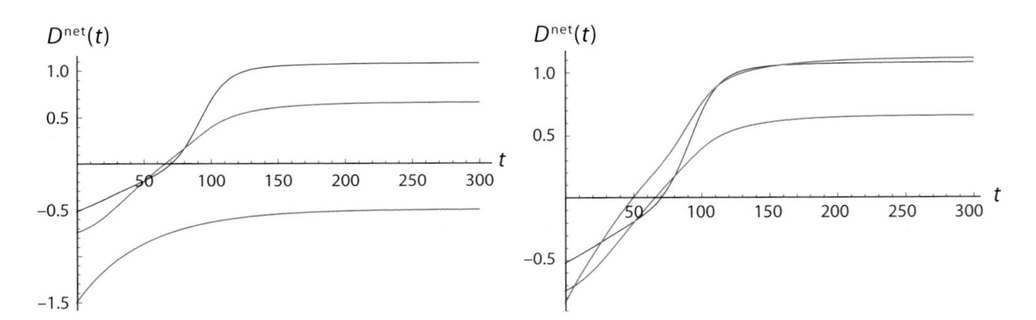

FIGURE 26. Heterogeneous Nonclassical Learners

militantly committed to the course . . . a course he would not have adopted in isolation.[108]

Succession Does Not Imply Imitation

Once again, this is not an example of behavioral imitation. Agent 3 is first to behave, so—ipso facto—cannot possibly be imitating anyone's behavior! But, in the model, neither is 1 or 2 imitating 3's *behavior*. Even those who act later are not imitating *behavior*. They cannot be. In the model, behavior is Boolean (0 or 1), and this binary variable doesn't appear as an independent variable in the model. Now, when Agent 3 exceeds threshold (crosses the horizontal axis), none of the others have done so, and their (binary)

[108] The extinction curve is shown in Appendix II. The extinction order is different from the acquisition order.

behavior is still 0. However, their affect—their solo disposition—was transmitted, and this sufficed to trip Agent 3's behavior. This is a fundamentally different picture than models of contagious *behavior*. This emerges even more starkly in the spatial agent-based version that follows.

Contra Great Man Theory

Is Agent 3 the "leader" of the group's ultimately unanimous lynch-mob behavior? By and large, we tend to assume that the *first to act* is the "leader," that he is the most committed to the rightness of his course, and the most passionate toward it, and that these qualities are what make him a "man of action" who inspires others. Here is a direct counterexample, suggesting that those who *act* first and are normally seen as "leaders" may have no distinguishing attribute save especial susceptibility to dispositional contagion. *The others do follow him in time, but it is not because he has inspired them, but because they have inspired him!* I certainly hope this is seen as a useful attack on the "great man" theory of history.

Tolstoy: The First Agent Modeler

Its most brilliant critic, and perhaps the first agent-based modeler, was Tolstoy. Tolstoy's *War and Peace* (1869; 1998 ed.) is, of course, a literary-historical masterpiece. But, its attack on the "great man" theory also foreshadows agent-based modeling. Tolstoy's perspective is entirely bottom-up and generative. It is a shame he did not live to see the advent of agent-based modeling, in which his view of history as an "emergent phenomenon"[109] is being realized computationally.

Tolstoy writes, "There are two sides in the life of every man, his individual life which is more free the more abstract its interests, and his elemental swarm-life[110] in which he inevitably obeys laws laid down for him." Tolstoy calls history ". . . the unconscious general, swarm-life of mankind," and abjures the historian's focus on "the great man" (p. 647). He writes, "The ancients have left us model heroic poems in which the heroes furnish the whole interest of the story, and we are still unable to accustom ourselves to the fact that for our epoch histories of that kind are meaningless. . . ."

In what I read as an incredibly prescient anticipation of agent-based generative modeling, Tolstoy continues, "To study the laws of history, we

[109] I use the term *emergent* advisedly. For an energetic critique of classical emergentism, and its contemporary offspring, see J. M. Epstein (1998, 2006).

[110] I thank my fluent colleague and eminent Russia scholar Clifford Gaddy for wonderful discussions of Tolstoy's usage here. He notes that Tolstoy's precise term is роевая, which is an adjectival form.

must completely change the subject of our observation, must leave aside kings, ministers, and generals, and study the common, infinitesimally small elements by which the masses are moved. No one can say in how far it is possible for man to advance in this way towards an understanding of the laws of history; but it is evident that only along that path does the possibility of discovering the laws of history lie; and that as yet not a millionth part as much mental effort has been applied in this direction by historians as has been devoted to describing the actions of various kings, commanders, and ministers, and propounding reflections of their own concerning these actions" (p. 881). "Only by taking an infinitesimally small unit for observation (the differential of history, that is, the individual tendencies of men) and attaining to the art of integrating them (that is, finding the sum of these infinitesimals) can we hope to arrive at the laws of history" (p. 880). But what is this but to "grow" history from the bottom up in an individual-based model? The agent-based model is the appropriate tool for studying the swarm-life of man.

While Agent 3 is the first to act in Figure 26, he is no more "a bold leader" than the fish at the head of the school. Neither is he following the direct orders of anyone else.

Swarmocracy

This is why conspiracy is less disturbing than decentralized swarm totalitarianism, and why conspiracy theory is correspondingly less profound than Tolstoy. Conspiracy theory attributes to the conspirators *the theorist's own analysis* of the situation and posits that the conspirators are cynically benefitting from that situation, as fully aware of it as the theorist. Sometimes this is certainly true, as in countless cases of industrial and political corruption. But, in a way, this is the easy case: chop off the head, and the conspiracy is done. If you are being attacked by a bear, one well-placed bullet solves the problem. But, that is not the swarm-life . . . if the enemy is the swarm of bees, no bullet works. *The swarm is resilient to local disruption and endogenously reassembles and perpetuates itself.* Those individuals—differentials of history—who find themselves at the head are not cynics; they actually believe the (often pernicious) mythology they propagate. That they obviously benefit from it does not make them cynics either. Indeed, the mythology itself permits them to infer their righteousness *from* their station, regardless of how their station may have been attained.

Summing up to here: we have constructed the central case, in mathematical form—the parable in which *Agent_Zero* initiates action through disposition contagion. In the discussion thus far the action has been violent, but we later offer many interpretations in which it is not.

Complex Contagion of Disposition

As noted earlier, three agents are required to depict majorities. But three are also required to depict the important phenomenon of *complex contagion*. In the case of many diseases, such as influenza or smallpox, a single exposure is sufficient to produce transmission. However, the transmission of attitudes, norms, fears, and other social signals may require multiple exposures, confirming exposures, if you will. Centola and Macy (2007) introduce the term *complex contagion* for these and demonstrate that complex contagion dynamics on networks differ fundamentally from simple contagion dynamics, indeed that certain basic results (e.g., the power of small world network structures) are not robust to the requirement of multiple (as few as two) exposures. Because they sum the weighted dispositions of other agents, the individuals in the present model can exhibit complex contagions. For example, the previous Blue agent (right panel of Figure 26) requires the confirming disposition of the Red agent to pull her over her threshold. Influenced only by the Green agent, she would not trip. But, reinforced by the third, she does.

Interdependent Dispositional Trajectories in 3-Space

The vector of dispositions $(D_1(t), D_2(t), D_3(t))$ from Figure 24 traces out a curve in 3-space, as illustrated in Figure 27.

Just as the thresholds defined a rectangle in the two-agent case, so the three agents' thresholds now define a cube, (a box parallelepiped) depicted in blue. There is unanimity to act when the group space curve is outside the threshold cube, as shown in Figure 28.

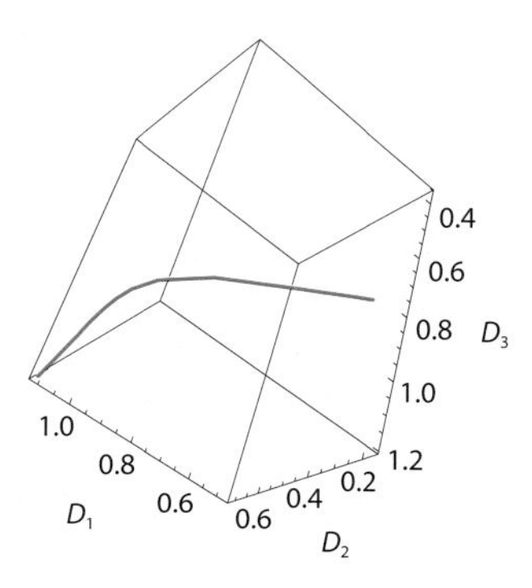

FIGURE 27. Group Dynamic as a Curve in 3-Space

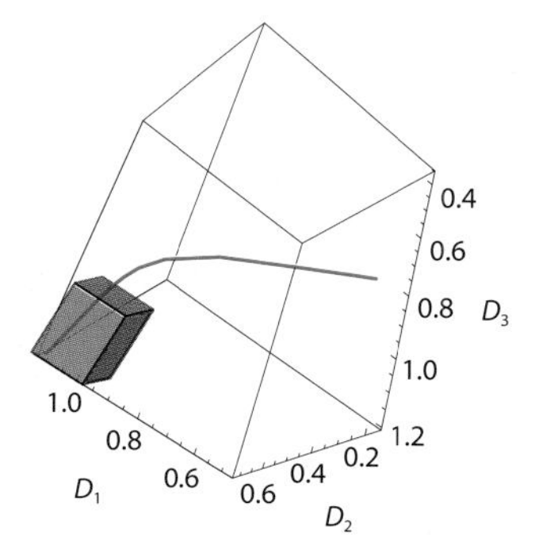

FIGURE 28. Unanimity When Group Trajectory Escapes Threshold Cube

A Mathematical Aside on Social Norms as Vector Fields

Obviously, these group trajectories (social movements, crowd behaviors) all take place in a social setting, which may be disapproving and resistant or supportive. These attitudes are local and can differ in different regions (neighborhoods of n-space in general). It is natural to think of these social norms as a resistant (or promotive) vector field through which the group trajectory is "working" to move.[111] How much social "work" is necessary to move from the origin to other positions in this milieu? That is the Line Integral of the trajectory in the Field, which we will compute and explore.

The examples thus far developed unfold in the degenerate "null vector field" of global indifference. The simplest significant departure from that would be the negative radial vector field directed toward the origin, itself representing the social state of absolute dispositional unanimity.

Negative Radial Moralities

The negative radial field assigns to each point in 3-space (R^3) a vector directed toward the origin, with magnitude equal to the point's distance from the origin. The farther from the origin the group members are, the stronger is the force directing them back to the origin. More relevantly, the more severely you violate a norm, the more severe is the pressure to return.

The radial vector field is $\mathbf{F} = (-x, -y, -z)$, where these entries are themselves real-valued scalar functions from R^3 to R^1 (Marsden and Tromba, 2011; first edition 1976). This is shown in Figure 29.

As a purely heuristic start, some norms do work roughly this way. If you commit two murders, you get two life sentences. Fines for speeding increase with the level by which you exceed the speed limit. In orthodox Jewish law, the Mishna states the Law of Lashes for drinking, shaving, and becoming unclean (failing to wash after proximity to a corpse). Speaking of the Nazir[112] the Mishna reads, "If they said to him, "Do not drink, do not drink, and he drank, he is liable to lashes *for each one*"[113] (Mishna Nezikin Makkot, Ch. 3).[114] That is, the punishment, the restoring force, is additive with the number of transgressions . . . the number of steps from the norm.

[111] For rigorous definitions of norms and models of their dynamics, see Axelrod (1997a), Young (1993), and J. M. Epstein (2006).

[112] One who has foresworn alcohol, shaving, and sexual intercourse.

[113] In fact, the maximum is 39 lashes. I thank David Broniatowski and Julia Chelen for Talmudic researches.

[114] Chapters 1 and 3 of the fifth tractate Makkot of the fourth order Nezikin in the Mishna. Retrieved from http://halakhah.com/pdf/nezikin/Makkoth.pdf.

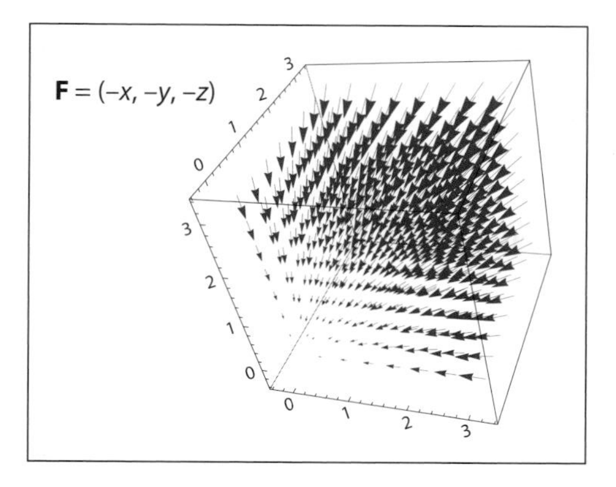

Figure 29. Negative Radial Vector Field

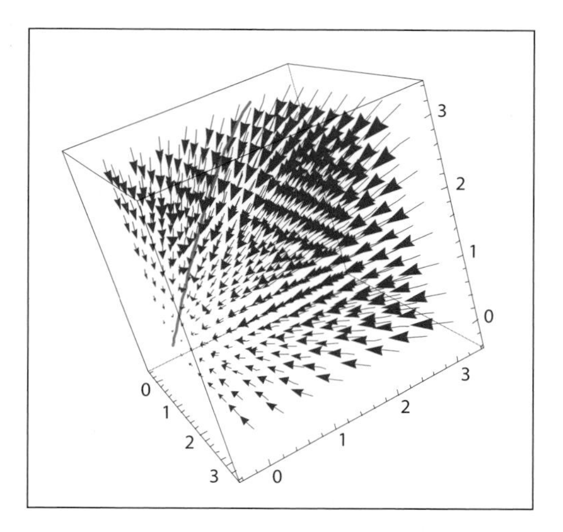

Figure 30. Group Trajectory in a Resistant Field

The same wording applies to the other violations: uncleanliness and shaving (Mishna Nezikin Makkot, Ch. 3). Adages such as "Let the punishment fit the crime" or "An eye for an eye; a tooth for a tooth" all express *negative radial moralities*, if you will.

Now if we imagine the (red) group trajectory as it moves through the negative radial norm field, the picture is as in Figure 30.

To even the most lapsed of physicists, this picture veritably cries out for a line integral. Specifically, if we imagine the radial field as force, then we can compute the work (as in physics) required for the group to move out any given trajectory subject to the field (Figure 31).

We have the affective space trajectory:

$\gamma(t) = (x_1(t), x_2(t), x_3(t))$. Now define a (resistant) Vector Field

$\mathbf{F} = (f_1(x,y,z), f_2(x,y,z), f_3(x,y,z))$. In this field, $\gamma(t)$ experiences

$\mathbf{F}(\gamma(t)) = (f_1(x_1(t), x_2(t), x_3(t)), f_2(x_1(t), x_2(t), x_3(t)), f_3(x_1(t), x_2(t), x_3(t)))$.

$\mathbf{F}(\gamma(t)) \in \Re^3$. So, we can form a dot product with any other 3-vector, like $\dot{\gamma}(t)$. Then,

$$\int_0^1 \mathbf{F}(\gamma(t)) \bullet \dot{\gamma}(t) dt$$

is the **Work** required to move from $\gamma(0)$ to $\gamma(1)$ in **F**.

FIGURE 31. Social Work as a Line Integral

If **F** is centrist (negative radial), the group's movement against it takes work, to a very surprising degree. If we just look at a straight-line trajectory directed away from the origin at 45 degrees to each axis, in a negative radial field (our earlier one multiplied by a scalar *a*, which was unity), and track the group's total work in moving from the origin to any trajectory point (κ, κ, κ), the work is calculated as follows.

$$\mathbf{F} = (-ax, -ay, -az) \quad [27]$$

$$\gamma(t) = (t, t, t) \quad [28]$$

$$\dot{\gamma}(t) = (1, 1, 1) \quad [29]$$

$$\mathbf{F}(\gamma(t)) \bullet \dot{\gamma}(t) = -3at \quad [30]$$

$$W = -\int_0^{\kappa} (3at) dt = -\frac{3}{2} a\kappa^2 \quad [31]$$

Very interestingly, the work is proportional to the *square* of κ. Hence, it is *very* hard to swim (far) against (even the unit negative radial) tide of convention. This applies only to the idealized negative radial field we've been exploring—a field that is irrotational (Curl $\mathbf{F} = 0$) but compressible (Div $\mathbf{F} < 0$). The true field, of course, is full of whirlpools, local equilibria, normative eddies, and moral turbulence. These would all be caused by inhomogeneities in the field—such as neighborhood norms, social movements, and so forth—all worthy topics for further modeling.

Algorithm

To review this little mathematical excursion, the (original) Rescorla-Wagner initial-value problem is solvable analytically:

$$\forall i, \frac{dv_i}{dt} = \alpha_i \beta_i (\lambda_i - v_i);\ v_i(0) = 0 \qquad [32]$$

$$v_i(t) = \lambda_i (1 - e^{-\alpha_i \beta_i t}). \qquad [33]$$

By substitution, so are the coupled net-of-threshold dispositions, $D_i^{\text{net}}(t)$.[115]

$$\begin{aligned} D_1(t) &= v_1(t) + P_1 + \omega_{21}(v_2(t) + P_2) + \omega_{31}(v_3(t) + P_3) - \tau_1 \\ D_2(t) &= v_2(t) + P_2 + \omega_{12}(v_1(t) + P_1) + \omega_{32}(v_3(t) + P_3) - \tau_2 \\ D_3(t) &= v_3(t) + P_3 + \omega_{13}(v_1(t) + P_1) + \omega_{23}(v_2(t) + P_2) - \tau_3 \end{aligned} \qquad [34]$$

These are the components of the space-curve [35]:

$$\boldsymbol{\gamma}(t) = (D_1(t), D_2(t), D_3(t)). \qquad [35]$$

Then we posited the centrist negative radial vector field (with the origin as its global attractor) $\mathbf{F} = a(-x, -y, -z)$. Denoting the space-curve's derivative as $\dot{\boldsymbol{\gamma}}(t) = (\dot{D}_1(t), \dot{D}_2(t), \dot{D}_3(t))$, we form the dot product

$$F(\boldsymbol{\gamma}(t)) \cdot \dot{\boldsymbol{\gamma}}(t) = -a \sum_{i=1}^{3} D_i(t) \dot{D}_i(t). \qquad [36]$$

Whence the work required is

$$W = -a \int_0^{\kappa} \left(\sum_{i=1}^{3} D_i(t) \dot{D}_i(t) \right) dt. \qquad [37]$$

The generalization to arbitrary n is direct. I have treated the resistant vector field of society as exogenous. *Ultimately, of course, it is made up of all the other dispositions—all Tolstoy's differentials of history—themselves the result of complex interactions.*

[115] We momentarily drop the superscript *net* to reduce visual clutter. Otherwise, these are identical to the formulae given in Figure 24.

Most discussions of learning cast trials as interactions between the experimenter and a subject. But, of course, interactions with other people are trials as well. There is no experimenter. In society, conditioning and extinction are underway concurrently, and in parallel.

This entire business of vector fields and their endogenization is a suitable topic for future research. Now, however, we return to the main plot of the book, without any embedding vector field—or, if you prefer, with only the null one.

Extinction of Majorities

Returning to the main plot, as discussed earlier, conditioning and extinction are not symmetrical. For example, recall that in Figure 26, agents crossed the threshold in the order Green, Red, Blue. But if we turn off all excitation trials at time 300, extinction (e.g., forgiveness, recalling our half-life discussion) occurs in a different order, namely, Red, Blue, Green, as shown in Figure 32. And the time to majority forgiveness is roughly half the time for majority excitation. Clearly, the space curve reenters the threshold cube (i.e., all curves in Figure 32 go negative) at approximately $t = 35$. Inside the cube, there is unanimous negative net disposition.

Posttraumatic Stress and Lambda

As mentioned earlier, the original Rescorla-Wagner model treats extinction by setting λ to zero when trials stop. But, as also noted, some people

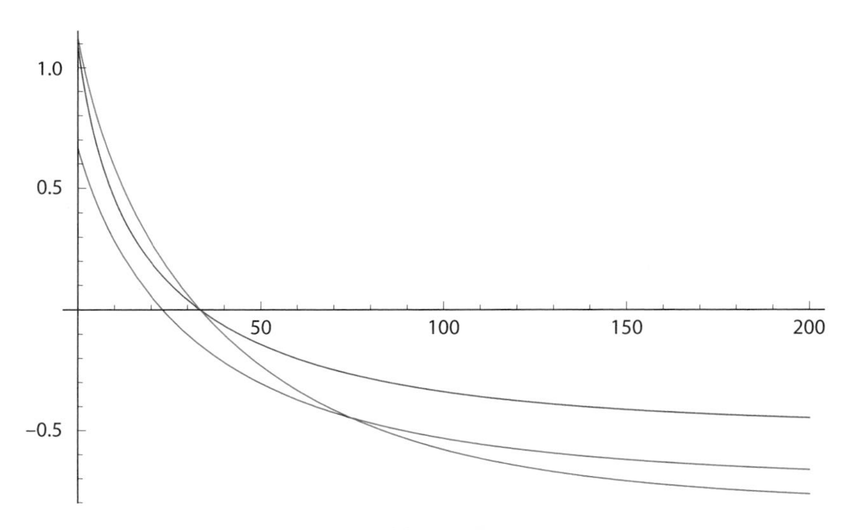

FIGURE 32. Majority Extinction

cannot set their λ to zero just because conditioning trials cease. They cannot free themselves of the negative association. This is often detrimental to the individual. But what is its effect on the group? With everything as in the previous simulation, if Agent 3 (Green) *alone* is traumatized and can cut her λ only to 0.95 rather than zero—as the other two can—*everyone's extinction trajectory is affected,* as shown in Figure 33 (as compared to Figure 32). Now, Agent 3 is never able to extinguish the trauma. Moreover, the other agents take longer to do so, and they are not able to recover the same (negative) level of pretrauma dispositions as in Figure 32. So, the effects are dramatic and enduring. The hypothesis, in other words, would be that, in close-knit groups (i.e., where weights are high) subject to trauma, a single individual's ongoing PTSD can delay and degrade *everyone's* recovery. We will pick up this theme again in the agent-based model when we "lesion" one of the agents and study the effect on others in her network.

On historical scales, to extinguish collective distrust or enmity borne of sustained brutality and injustice (repeated aversive trials) may take more than their cessation, or even the good will of a new generation. Also, the negative association is easily reinstated by a recurrence of aversive experience (Rescorla and Heth, 1975; Norrholm et al., 2006; Bouton, 2004). In fact, the reinstatement can be more dramatic than on the initial trials. The wife cheats. The couple goes through extensive counseling. The husband's trust is restored. If, after all that, the wife cheats again, it can be more surprising and salient than the first time. Likewise, revelations of racial health disparities today are amplified by the long shadow of Tuskegee.

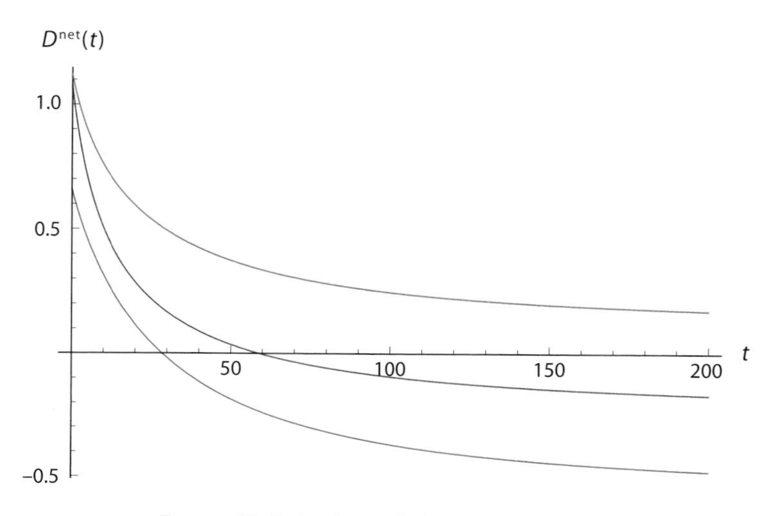

Figure 33. Extinction with One PTSD Individual

I.4. INTERIM CONCLUSIONS

The purely mathematical model, albeit highly abstract and very simple, does capture some core phenomena of interest. Behavior (the binary action adopted) is the result of affective, deliberative, and social modules. These are combined to form a total disposition (D^{tot}), which is compared to a threshold (τ). If D^{tot} exceeds τ,[116] the agent acts; otherwise, she does not. Social networks are the medium over which social influence flows, but centrally, agents do not imitate one another's binary behavior,[117] but *dispositions*. Solo dispositions (each a sum of affect and probability judgments) are contagious via weights; if these conspire to produce total dispositions exceeding thresholds, observable behavior results. But the first agent to behave—the "leader"—may actually not be either the most passionate or the most convinced by evidence but, rather, may simply be the most susceptible to dispositional contagion, for which some current research support was offered. Asymmetries between fear acquisition and its extinction were noted, and the potentially large social effects of heterogeneous minimum λs (as a representation of PTSD) were briefly explored.

Mainly, we generated the central case: *the agent who initiates the group's behavior despite starting with the lowest disposition, with no initial emotional inclination ($V = 0$), no evidence ($P = 0$), the same threshold (τ) as all others, and no orders from above.*

In fact, the preceding mathematical example begins with *negative* net disposition. The agent is actually *dis*inclined toward the act of interest, be it violence, vaccine refusal, binge eating, or dumping toxic assets. But, through dispositional contagion, she is the first to act.

Revealing as it is, the simple coupled differential equation *Agent_Zero* model developed previously is quite unrealistic in three respects. First, the probabilities P are exogenous and fixed, rather than updating with recent direct experience or influence of others. Second, training trials are assumed to be arriving continuously, rather than in a discrete stochastic process. Third, the model is nonspatial. Agent-based modeling is a powerful way to include these features.

[116]Equivalently, if $D^{net} = D^{tot} - \tau$ is positive.

[117]Technically, they do not register this.

PART II

Agent-Based Computational Model

In agent modeling, we essentially build artificial societies of software individuals who can interact directly with one another and with their environment according to simple behavioral rules. On agent-based modeling in general, see, for example, J. M. Epstein and Axtell (1996), Axelrod (1997a), Resnick (1994), J. M. Epstein (2006), Tesfatsion and Judd (2006), Miller and Page (2007), and the large literature cited in these works.[118]

I developed this model in *NetLogo* 5.0. Source Code for the canonical[119] Parable 1 run is given in Appendix IV. A table of parameter values for every run is also provided. As earlier noted, all movies are posted on the book's Princeton University Press Website. Interactive Applets for each movie run are provided there as well. The Applets allow the user to alter various assumptions with "sliders," movable bars on the Interface. These user-adjustable parameters include the attack rate, search radius, extinction rate, memory length, and damage radius, for example. This offers nonprogrammers an extensive basis for experimentation with the model. For programmers, the Applets also include the full source code for every run. Hence, all results are certainly replicable. However, the English-language exposition that follows is meant to be sufficient to permit replication by reasonably adept programmers (who are also good readers).

Replicability

Apropos of this, I am not sure replicability is an attribute of models proper. Leaving aside the case of authors who are literally pretending to have a model, one could always "replicate" model output by running the same model on the same inputs. So, when a person says a model was not

[118] A nice "Guide to Newcomers" is available in Axelrod and Tesfatsion, Appendix A of Tesfatsion and Judd, eds., *Handbook of Computational Economics: Agent-Based Computational Economics, Volume 2*. Among the closest things to a textbook on agent modeling is Railsback and Grimm (Princeton 2011). The best hands-on way to get started is to do the three excellent agent-based modeling tutorials that download with *NetLogo* (http://ccl.northwestern.edu/netlogo/).

[119] By canonical, I mean simply the base model for this development.

replicable from some article, they are really asserting that the author's *English-language exposition* of the algorithm was insufficient to permit a reimplementation by that particular reader. If so, it would appear to measure the author's facility in English—or the reader's lack thereof—but it has nothing to do with the actual computer program or mathematical equations, which—if provided, as here—are replicable ipso facto. In any event, such ambiguities as may arise can be resolved by reference to the code provided in Appendix III and on the book's Princeton University Press Website.

Present Interpretation

Later, I will offer a number of alternative interpretations of the model in the fields of health behavior, economics, network science, and law. But for expository purposes, we imagine a conflict, indeed a guerilla war like Vietnam, Afghanistan, or Iraq. As discussed in the Introduction, events transpire on a 2-D population of contiguous yellow patches, each of which represents an indigenous agent. Specifically, we imagine that a single stationary indigenous agent occupies each patch. This expository grid is 33 by 33 (the default *NetLogo* dimensions), so there are 1089 Yellow agents.

These indigenous patch-agents do not move. They have two possible states: inactive and active. At any point in time, they occupy only one of these states. Inactive agents are yellow. I have given them slightly different shades of yellow just so they are visually distinguishable squares, as shown in Figure 34. Active agents are orange. These agents activate randomly, at a rate (the attack rate) adjustable by the user. They will be discussed shortly. The three Blue agents represent occupying forces and are of the full *Agent_Zero* type. They are mobile. Every cycle through the (randomized) agent list, the agent adopts a random heading and takes one step in that direction. So, they do not jump to random distant sites but move to random neighboring ones. They execute a 2-D random walk, in short. It is not a perfect mixing, or mass-action kinetics, process. The space is a finite bounded square lattice.[120] These Blue "rovers" are connected to one another bidirectionally, as indicated by the two-headed arrows shown in Figure 34. (Hence, there are in-degree and out-degree distributions, and so forth). This exactly parallels the mathematical networks developed above. Some rovers give high weight to other rovers; some do not (see the Appendix IV table, or *NetLogo* Code in Appendix III for the values employed).

[120] A torus topology is readily available in *NetLogo* but would have been visually confusing for most of the runs explored here.

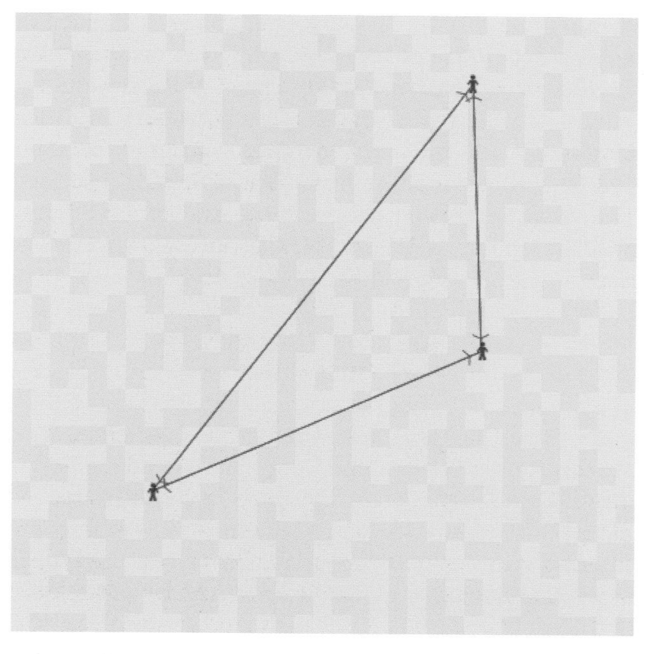

FIGURE 34. Indigenous Population (Stationary Yellow Squares) and Occupying Rovers (Mobile Blue Agents) [Movie 1]

Minimalism

I have hand-coded three agents to ensure complete control over specifics in the smallest possible model that can exhibit majorities. For large *n*, one would of course initialize the agents with random weights and other parameters drawn from distributions. The agents and their connections are shown in Figure 34.

Movie 1 (on the Princeton University Press Website) simply shows the three agents in random motion connected to one another.

Now, let us posit a distinguished region of the space—in this version, the northeast quadrant—where yellow patches "activate" at a user-specified random rate (a global constant, implemented as a user-adjustable slider in the *NetLogo* interface).[121] Think of these agent activations as insurgent attacks. These explosions are shown as orange patches in Figure 35. I make no

[121] *NetLogo* offers a wide variety of distributions from which to draw random numbers. Here, $U(0, 1)$, the uniform distribution on the unit interval, is used.

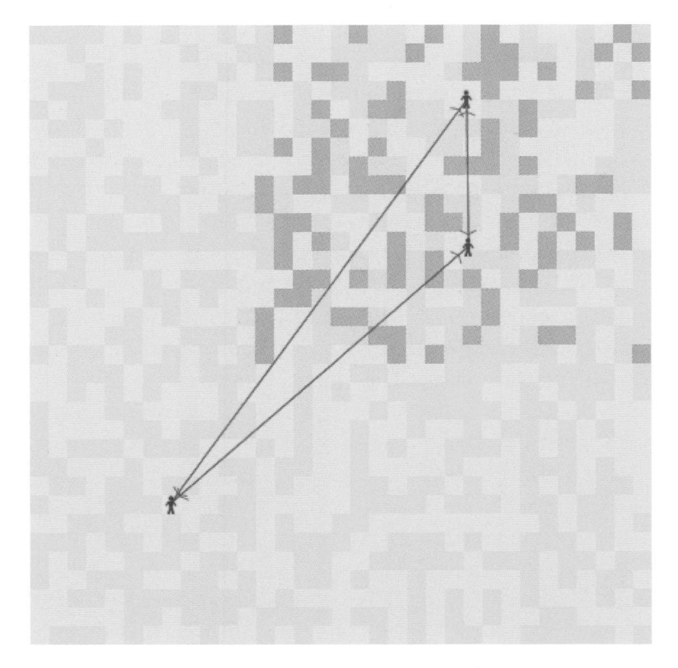

FIGURE 35. Adverse Event Activations [Movie 2]

assumptions whatsoever as to the comparative legitimacy of occupying or insurgent agents. **Movie 2** shows these, with fixed Blue agents.

It is central to distinguish between a Blue agent's separate affective, "rational," and social components and to understand how they are combined to form the agent's overall disposition in the wake of attack. As before, once this disposition is formed, it is compared to the agent's threshold. If the overall disposition exceeds threshold, action is taken; otherwise it is not. In this interpretation, *action is the destruction of indigenous (Yellow) agents within some user-specified damage radius* (again, a slider in the *NetLogo* Interface). Now each component is described.

Affective Component

These orange activations (explosions) are the conditioning trials for the Blue agents. When a Blue agent "steps on" an orange patch, he updates his affect through the generalized Rescorla-Wagner equations.[122] Learning rate parameters (the α's and β's), limiting values for associative strength (λ), and

[122] *Update-affect* is the relevant *NetLogo* code block. See Appendix III. My code extends Rescorla-Wagner in allowing extinction rates different than the classical model, which, of course, is an available setting.

the exponent (δ) all affect individual learning curves and can vary across agents. An extinction rate is applied at every iteration except those in which an active patch is encountered. So, hostile affect toward the indigenous population evaporates at a user-specified extinction rate in the absence of local attacks.[123] This extinction rate can be zero since extinction, as noted earlier, is by no means assured simply by cessation of trials. This is the *affective* component of the Blue agents' disposition.

A suitable extension would be to include the well-established contextual conditioning that Blue agents would presumably undergo in the course of their spatial movements. They would come to associate the northeast quadrant itself with danger, and this would amplify the estimates made purely from event sampling. The hippocampus is central to this well-established contextual conditioning in space. In animal models, Knierim (2009) and Knierim and McNaughton (2001) have used "multi-electrode arrays to record the extracellular action potentials from scores of well-isolated hippocampal neurons in freely moving rats. These neurons have the fascinating property of being selectively active when the rat occupies restricted locations in its environment. They are termed *place cells*, and it has been suggested that these cells form a cognitive map of the environment (O'Keefe and Nadel, 1978). The animal uses this map to navigate efficiently in its environment and to learn and remember important locations" (from Knierim Research Page, Johns Hopkins Mind/Brain Institute site). *Agent_Zero* agents do not have a mental map of the area and condition only on the event stream, not also on position, though an Agent 1.0 could certainly have this endowment.

"Rational" Component

Turning to the evidentiary/ratiocinative component, Blue agents have a *spatial sampling radius* (which can be heterogeneous but is also a slider in the *NetLogo* Interface), within which they conduct local sampling of *the landscape*,[124] here interpreted as an indigenous population.[125] As discussed earlier, they estimate the probability that an agent is a hostile agent (e.g., the probability that an agent is a terrorist given that he is Muslim) by computing the relative frequency of orange patches within their sampling radius.[126]

[123] Properly speaking, this extinction-rate slider is a multiplier. If it is 0, there is no extinction. If it is 1.0, we obtain classical Rescorla-Wagner extinction curves. Typically, we use a value in the interval (0,1). So, this is a second extension of the original model (beyond *S*-curve learning), permitting yet another type of flexibility.

[124] Hence the adjective "spatial."

[125] Later, we will interpret the set as a space of financial assets, a family of vaccines, or opportunities for unhealthy eating, over which a *local relative frequency* is being computed and updated.

[126] This is the number of orange patches over total patches within the spatial sampling radius.

Obviously, this probability estimator exhibits *sample selection error*—the local ratio may be a poor estimator of the global one.

Some readers may feel that this simple computation is putting "reason" at an unrealistic disadvantage to "the passions." In fact, while this sample estimate is crude statistically, its computation is remarkably sophisticated cognitively. Indeed, this imputes to the Blue agents more cognitive capacity than untrained humans possess. In *The Mathematical Brain,* Butterworth (1999) makes a powerful argument that among our innate universal endowments is a *number module*, giving us the capacity to make crude numerosity judgments; and he provides evidence that the parietal lobes are centrally implicated. So, just to be shamelessly phrenological, while *Agent_Zero* walks into an ambush, his amygdala is activated and so he registers fear, but his number module is also making a very crude frequency judgment: *enemy/ total.* Butterworth argues that even this simple relative frequency is very hard for humans to compute, which suggests a neural basis for one of the best-documented biases in all of psychology: base rate neglect (Kahneman and Tversky, 1973; Tversky and Kahneman, 1982). As he puts it, "we ignore base rates because we ignore rates" (Butterworth, 1999). So, simple as it seems, *Agent_Zero*'s computation of a local ratio is far from trivial.

Another factor that would corrupt the Blue agent's estimate of the actual local ratio (itself a biased estimator of the global one) is the specific areal pattern in which the activations present themselves. In experiments, human subjects are quickly shown two spatial arrangements of dots: one has them spread over a wide area, and the other has them tightly packed. We will judge the former pattern to be the more numerous (Krueger, 1972, 1982). One can imagine how this areal bias might have conferred a selective advantage—we are more vulnerable (and so more alert) if surrounded by predators than if they are all clustered within our vision (giving us more escape routes).

It also happens that, even if our dots occupy the same total area, random patterns (as in this model) are typically overcounted as against regular ones, again perhaps because unpredictable predator patterns are harder to anticipate and evade than regular ones.

Related mechanisms may explain why we involuntarily complete patterns like those shown in Figure 36. The seminal example is the Kanizsa triangle (Figure 36A), after Italian psychologist Gaetano Kanizsa (1955).

This "phantom edge phenomenon" (seeing an outline that is not actually there) is due to what neuropsychologists call the "T-effect."

> Groups of neural cells see breaks in lines or shapes, and if given no further input, will assume that there is a figure in front of the lines. Scientists believe that this happens because the brain has been trained to view the break in lines as an object that could pose a potential threat. With lack of additional information, the brain errs on the side of safety and perceives the space as an object. The circle is the

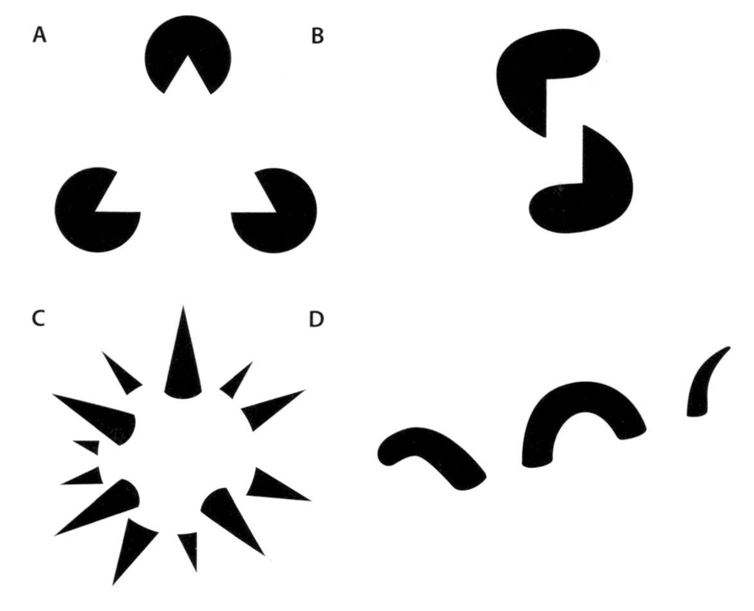

Figure 36. Phantom Edges

> most simple and symmetrical object, so the mind usually sees a circle unless active effort is made to see an alternate shape. This illusion is an example of reification or the *constructive* or *generative* aspect of perception, by which the experienced percept contains more explicit spatial information than the sensory stimulus on which it is based (Ehrenstein illusion, n.d.).

This is yet another source of potential Blue agent "threat inflation" that we shall ignore. I would say that propaganda generally—in "completing" political patterns that aren't there or inviting their completion—traffics on this same apparatus. Indeed, the entire art of propaganda is to offer as little of the picture as possible, leaving it to the audience to "fill in the blanks" opened by vague outlines of subversive "others."

Finally, the base model treats the affective and statistical estimates as independent when they are almost certainly entangled. There is interesting experimental work on the classification (as hostile or peaceful) of inconclusive data, specifically under circumstances of threat (Baranski & Petrusic, 2010). Affect, in other words, colors one's probability judgments, particularly in settings of the sort we have posited.[127] I introduce this in an

[127] An anxiety-provoking context without question (Behrens et al., 2007).

extension of Part III. By contrast, in the basic model, these are superposed but decoupled—neither is a mathematical function of the other.

In sum, Blue agents simply compute the relative frequency of orange patches within their spatial sampling radius to estimate the likelihood that patches are immanently violent. This initially appears to be a very crude algorithm. In fact, it would probably be way beyond most humans, particularly in the stressful circumstances of interest here. But, since we want to give reason a "fighting chance" against passion, we'll start here. So, we now have an elementary type of bounded rationality, in addition to a simple representation of affect.

Social Component

The third *Agent_Zero* ingredient is social. At any time *t*, the total disposition of each agent is the sum of her affect, $V(t)$, and her local probability estimate, now a function of time, $P(t)$, plus the sum of each other agent's weighted solo disposition (each the sum of their own V and P), all minus her threshold.[128] Unless otherwise noted, the term *disposition* will denote *net disposition* in all *NetLogo* graphical output.

Sampling and Dispositional Radii Mathematically Independent

It is important to reiterate that the mechanism of influence in the model is not behavioral imitation, even if agents are within the narrow spatial

[128] Lest there be any replicative or other confusion, the agent source code and *NetLogo* graphical output use the name *disposition* for *net disposition*. This should occasion no confusion. The *NetLogo* code block (see Appendix III, p. 218) governing this calculation is:

```
to update-disposition
  ask turtle 0 [
    set disposition affect + probability + [weight] of red-link 1 0 * ([affect] of turtle 1 + [probability]
    of turtle 1) + [weight] of red-link 2 0 * ([affect] of turtle 2 + [probability] of turtle 2) − threshold]
  ask turtle 1 [
    set disposition affect + probability + [weight] of red-link 0 1 * ([affect] of turtle 0 + [probability]
    of turtle 0) + [weight] of red-link 2 1 * ([affect] of turtle 2 + [probability] of turtle 2) − threshold]
  ask turtle 2 [
    set disposition affect + probability + [weight] of red-link 0 2 * ([affect] of turtle 0 + [probability]
    of turtle 0) + [weight] of red-link 1 2 * ([affect] of turtle 1 + [probability] of turtle 1) − threshold]
end
```

Terms could be collected in a variety of ways, all equivalent computationally but different conceptually. This form seems expeditious for expository purposes. *NetLogo*'s name for a generic agent is "turtle." I choose to imagine that this is in honor of a famous exchange between Bertrand Russell and an audience member who told Russell that the earth was supported on the back of a great turtle. Russell asked, 'And what, pray tell, is supporting *that* turtle?' The answer was immediate. "Oh, another turtle . . . it's turtles all the way down."

sampling radius of one another. In Part III, an extension offers a way to introduce this distinction. But we do not use it in the main development. Agents can influence each other (have dispositional weight) at *any* range, by a large variety of avenues (e.g., auditory and textual social media), and the binary actions of other agents can alter the landscape (by destroying sites), which can affect one's frequency calculation. But binary action proper is not registered or, therefore, imitated. The spatial sampling radius is typically a cluster of contiguous sites on the landscape proper, such as a Von Neumann neighborhood. This spatial sampling radius is bounded and landscape specific. The radius of dispositional contagion is neither; the two are mathematically independent.[129]

Action

If overall disposition is greater than the threshold, action is taken: the agent destroys all patches within a user-specified damage radius (another slider).[130] Destroyed patches (indigenous agents) are colored *very* dark (i.e., blood) red and cannot be active (they are dead).

Pseudocode

So, for each Blue agent, the algorithm (pseudocode) is as follows:

```
Compute own affect (with orange explosions as conditioning trials);
Compute own local probability (relative frequency of orange within
   spatial sampling radius);
For each other agent in network
   Compute the weighted solo disposition;
Add the above-computed numbers;
Subtract own threshold;
If the result is positive, Act; otherwise don't;
Apply own extinction rate to own affect;
Move;
Repeat.
```

[129] As noted, agents can be in the same dispositional network even if they are not within one another's spatial sample radius. In such cases, communication (and dispositional contagion) could be by voice, by text message, by iPhone, by field radio, or other social media. Below we offer an extension allowing one to change weights step-functionally when others enter (or exit) one's sampling radius. We do not exploit that in the main exposition.

[130] Later, I endogenize this radius as a function of affect.

What Is Time?

Finally, it is worth noting exactly what we mean by "time" in this model. In the agent model (as against the continuous-time differential equations) time is discrete. Here, time advances by one unit with every complete updating cycle of every agent and every patch.[131]

II.1. COMPUTATIONAL PARABLES

Science begins as parable, and ends as probability.[132] As this is a very young science, the runs that follow are closer to parables than to mature scientific claims of any sort. They arguably qualify as explanatory candidates in a broad sense, in that they *generate* certain qualitative behaviors (see J. M. Epstein, 2006). But they are computational parables—fables if you prefer. Of course, some fables endure.

Parable 1: The Slaughter of Innocents through Dispositional Contagion

For the base case run of the agent model, we will immobilize one of the agents. Call him Agent 0. *Netlogo* begins subscripting agents from 0, so this numbering assures consistency with the code provided. But "*Agent_Zero*" is the name of a class, while "Agent 0" is the name of an individual

[131] Technically, time is measured in *ticks*, a reserved word in *NetLogo*. In this model I advance *ticks* by 1 with each cycle through the *NetLogo* "go" routine, which corresponds to *main* in C or C++. In this case, the full "go" code is a follows:

```
to go
  if ticks > = maximum-stopping-time [stop]
  move-turtles
  activate-patches
  update-event_count
  update-affect
  update-probability
  update-disposition
  take-action
  deactivate-patches
  do-plots1
  do-plots2
  do-plots3
  tick
end
```

[132] For example, the primordial fire god parables have been displaced by the uncertainty principles of quantum mechanics.

instance of that class. Lest any confusion arise, all the agents are of the general *Agent_Zero* type, just as all the diverse actors in a classical economic model would be of the *homo economicus* type (with different parameters, for instance). Agent 0 will be stationary in the southwestern quadrant of the landscape. The other two agents, Agent 1 and Agent 2, will execute random walks on the landscape but will begin in the hostile northeast quadrant. Agents can sample only the four sites to their immediate north, south, east, and west—their Von Neumann neighborhoods. All agents update their affects and local probability estimates, with dispositions updating over the fully connected network. For the Base Case, there is no affective extinction.

Crucially, Agent 0 is never attacked. As he is subjected to no direct conditioning trials, his immediate direct *affect is zero throughout.* Because he encounters no orange attack events, his estimate of probability (hostile given indigenous) is also *zero throughout.* Yet, he wipes out a "village"! How? As shown, the two rovers are encountering attacks (orange events). They are updating both their affect and their local estimate of the attack probability. When their total disposition values exceed their thresholds, they retaliate within their destructive radius (here equal to their sample radius). Destroyed sites are dark red, as shown in the left frame of Figure 37. Their destruction and the escalation of their affects and probabilities continue. At all times, Agent 0 is weighting these (i.e., their solo dispositions) and adding them to his own destructive disposition.

Finally, these push him over his own threshold and he wipes out innocents, despite having a sample probability of zero, and no direct emotional grievance

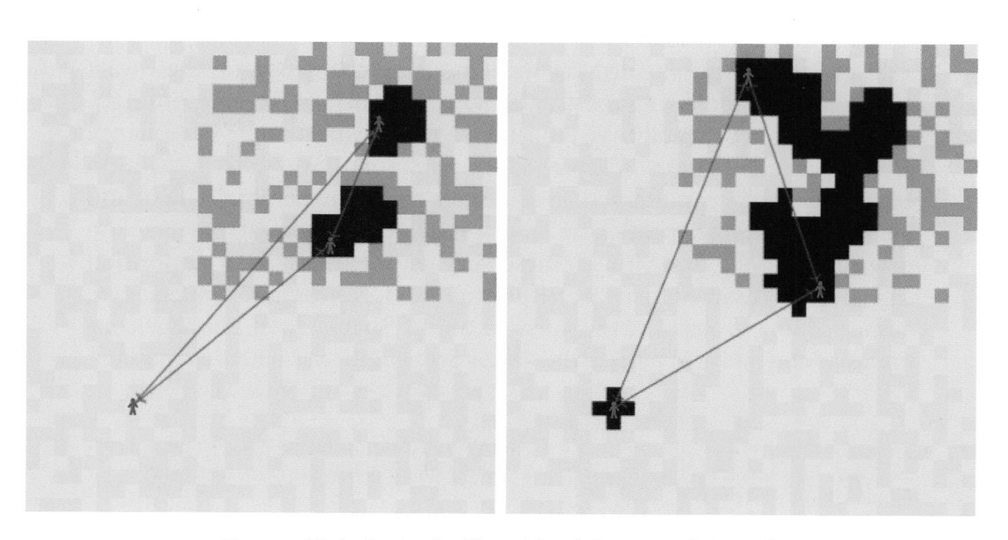

FIGURE 37. Activation by Dispositional Contagion [Movie 3]

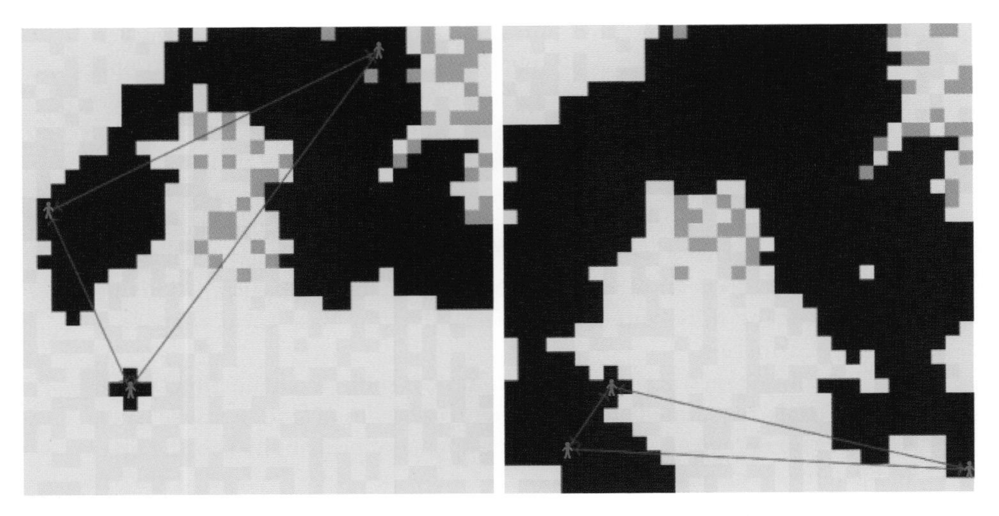

Figure 38. Slaughter of Innocents Continues [Movie 3]

against the population, as depicted in the right frame.[133] *Also, Agent 0 is not imitating the destructive behavior of either other agent.*

In this particular case, he cannot even observe their destruction of the landscape because "vision"—the sample radius—is set to one patch in each direction.[134] With no extinction of affect, the mobile rovers go on to wreak vast destruction in regions that have never done them harm either, as shown in the two frames from **Movie 3** shown in Figure 38.

Specifically, having wiped out many of the insurgents (in the northwest quadrant) and having now drifted out of that quadrant, the rovers are, in fact, encountering mostly yellow (innocent) patches. Accordingly, their estimated probability of a hostile patch (the local relative frequency of orange) falls to zero. Yet, without any evaporation of affect—with no extinction of the conditioned affect—their dispositions remain high, and the killing continues with no direct empirical (observational) basis and no new conditioning trials. This is shown in the time series of net disposition and probability in Figure 39. Disposition and destruction remain high, despite falling probability for Agents 1 and 2. That is, their rampage continues as all empirical basis for it—their probability estimate—evaporates.

[133] One might well say that Agent_Zero betrays himself in that his solo disposition is below his threshold, whereas his total (in the group) disposition exceeds it. In the nomenclature of the Introduction, $D^{tot} > \tau > D^{solo}$.

[134] As noted earlier, emotion and disposition can be communicated by numerous routes beyond immediate vision.

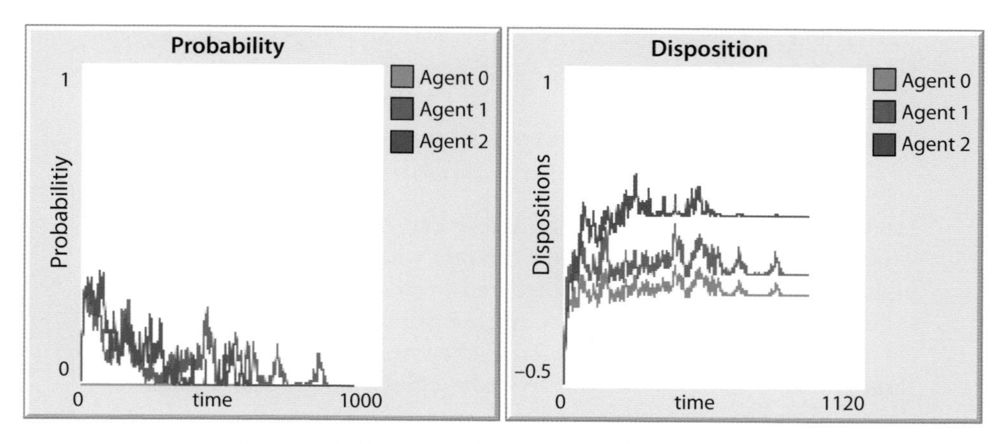

FIGURE 39. Rising Disposition Despite Falling Probability

This behavior is consistent with seminal laboratory psychology work of Zillmann et al. (1975).

Zillmann's Experiment

The article describing this experiment in detail is aptly entitled "Irrelevance of Mitigating Circumstances in Retaliatory Behavior at High Levels of Excitation." In sum, Zillmann et al. (1975, p. 282) showed that "Under conditions of moderate arousal, mitigating circumstances were found to reduce retaliation. In contrast, these circumstances failed to exert any appreciable effect on retaliation under conditions of extreme arousal." Specifically, "the cognitively mediated inhibition of retaliatory behavior is impaired at high levels of sympathetic arousal and anger." These conditions of affective arousal are certainly met, and agent behavior is entirely consistent with Zillmann's result.

Again, Agent 0's probability (the green curve of the left panel of Figure 39) is zero throughout. He would never have acted alone. And, he would never have acted even in a model of behavioral imitation, because he literally cannot "see" the others, and he need not, if they are in other forms of communication, such as auditory and social media.[135] The entire *NetLogo* Code for this parable is provided in Appendix III and again on the Princeton

[135] It is important not to muddy the distinction between the spatial sampling radius and the distance over which dispositional contagion may occur. The two are completely independent in the model. Weights do not increase with spatial proximity or shrink with distance. An extension allowing this is offered under Future Research.

University Press *Agent_Zero* Website. This is a disturbing run,[136] but it is not yet our canonical central case, because Agent 0 does not act *first.*

Parable 2: Agent_Zero *Initiates: Leadership as Susceptibility to Dispositional Contagion*

Having developed all this apparatus, we can now generate that case, in which the first agent to act is not the one with the highest affect or the highest empirical estimate of indigenous hostility. Indeed, Agent 0 (again stationary) is subject to no direct aversive stimuli (orange explosions), so his individual (i.e., directly stimulated) affect and probability are both zero throughout, as shown in the corresponding plots of Figure 40. By contrast, the mobile rovers are subject to attacks, are accumulating affect, and are increasing their estimates of the probability that an indigenous patch is hostile (that a random patch will turn orange). All thresholds are equal at 0.5,[137] but neither rover's disposition exceeds this, so neither of them acts. Through their weights, however, their dispositions elevate Agent 0's to the highest of levels (see disposition plot), which exceeds the common threshold first. So, he is the first to act, as shown in the Figure 40 screen shot and **Movie 4**.

This is the situation I aimed to generate: *The agent at the front of the lynch mob has no particular grievance V or evidence P, and left to his own devices would never act. Notice that this is not "the banality of evil." Agent 0 is not "just following orders," because none are issued. And he is not imitating the behavior of others, because he is the first to behave!* The deeper point, as emphasized throughout, is that no agent is imitating *the behavior* of others, regardless of the order in which they activate. Thus, the model can generate important group dynamics without recourse to the copying of behavior (which is a binary variable that doesn't enter into the disposition calculus). Action (i.e., behavior) occurs if total disposition exceeds threshold, which occurs first for Agent 0.[138]

[136] Even 10% extinction alters this considerably. A little forgiveness, or counter-learning, can go a long way.

[137] Notice that this makes Agent 0's solo net disposition negative in fact, since it is v (here 0) plus P (here 0) minus τ (here 0.5). The others' dispositions begin negative but rise quickly with aversive stimulus. Notice also that we do not arrange the activation order by giving agents different thresholds.

[138] There is an asymmetry in the model as developed to this point. In defining the binary action to be X (equal to 1) rather than not-X, one induces a reference direction. In the cases just described it is positive. One acts when one exceeds the threshold, not when one drops below it. As we see, the solo dispositions of other agents can indeed move one's net disposition in a positive direction (e.g., from negative to positive). But, since solo dispositions are nonnegative, they cannot move net disposition in a negative direction, that is in a direction contrary to the reference direction. To permit this, various mechanisms present themselves. One is threshold imputation, which I introduce in Part III, to replicate the Darley-Latane experiment. Another would be to introduce negative

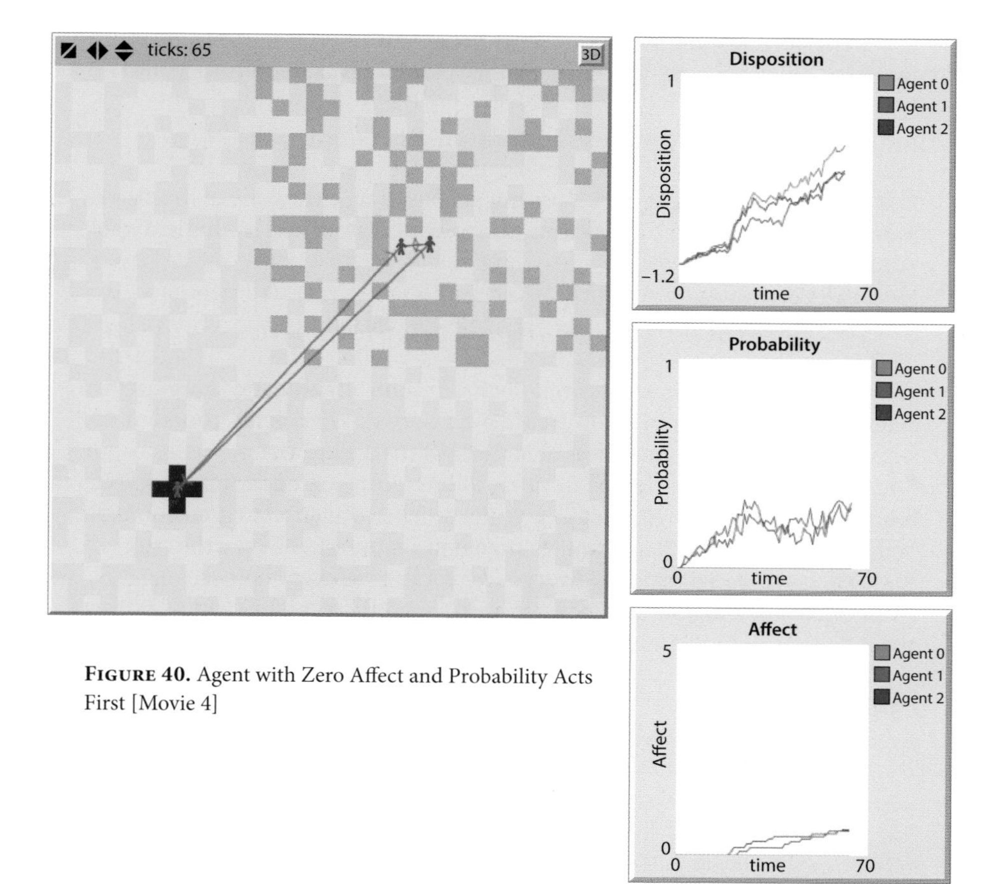

FIGURE 40. Agent with Zero Affect and Probability Acts First [Movie 4]

So, is Agent 0 a "leader," or is he simply the most susceptible to dispositional contagion? Which is the more compelling picture: the "great man" theory, or merely the susceptible one?[139]

This is Tolstoy's *swarm-life of man* in its most virulent form. Speaking of Bonaparte, Tolstoy writes:

> Though Napoleon at that time, in 1812, was more convinced than ever that it depended on him . . . he had never been so much in the grip of inevitable laws, which compelled him, while thinking he was acting on his own volition, to perform for the swarm-life—that is to say for history—whatever had to be performed." (*War and Peace*, p. 648)

weights. A third variation, for which I thank Julia Chelen, would be to have agents assign weight to the *average* of others' Vs and/or Ps. I also thank Jon Parker for discussions of this issue.

[139] See Tolstoy (1869; 1998 ed.)

"To perform for the swarm-life. . . ." What a phrase! And this is the sense in which Tolstoy wrote, "A king is history's slave" (*War and Peace*, p. 647).

Complex Contagion Revisited

It is worth noting that Agent 0 does not act based on either one of the others alone. Here, he requires the swarm, the weighted sum, and multiple dispositional exposures, to go.

Run 3. Information Cuts Both Ways

In the runs thus far, the agents' spatial "vision" (landscape sampling radius) has been limited to a von Neumann (N, S, E, W) neighborhood of radius 4 patches. What is the effect of increasing the agents' vision? Let us begin with Agent 0 in his usual fixed position in the southwest quadrant, which we assume to be peaceful. Now let us give the other two agents fixed positions as well, but in the violent quadrant, as shown in Figure 41. What is the effect of increasing everyone's vision? More peace? More violence? Neither?

Let's consider Agent 0. His vision is his spatial sampling radius. As this extends into the red zone, he is seeing more violence. Hence, his estimate

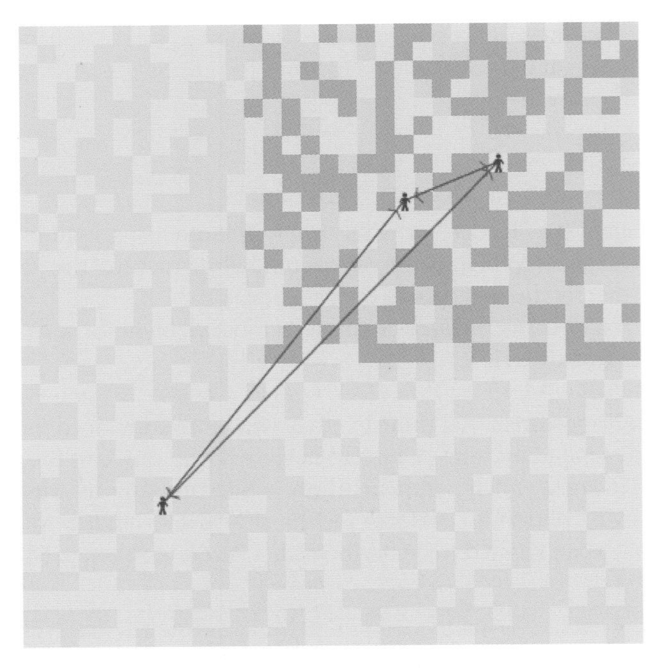

Figure 41. Fixed Agent Positions

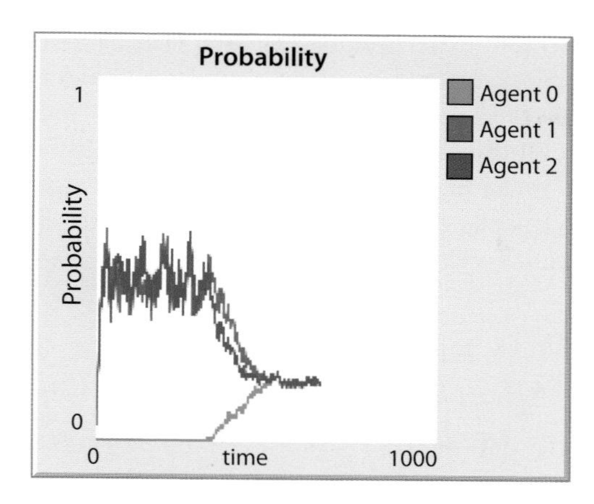

Figure 42. Probabilities Converge

that a random patch is violent grows, as will his violent disposition. The agents in fixed positions in the red zone, however, have the reverse experience. Rather than seeing more violence as their vision grows, they see more yellow—peace! Accordingly, their probability estimate falls. Finally, when vision increases to the point where they can all see the entire landscape, their probability estimates converge, because their samples are now identical. Notice that the sample selection biases were very great at the low-vision outset, with Agent 0 underestimating—and the others overestimating—the global probability. Now they converge on the correct global probability, as shown in Figure 42, where I simply increased the sampling radius mid-run with the program's slider.

This is an example of how sensitivities can be explored "on the fly" (mid-run) in *NetLogo*, which readers are invited to do using the interactive Applets posted on the book's Princeton University Press Website.

Heterogeneous Vision

In this experiment (and in this variant of the model), vision was the same for all agents. Again, I am using the term *vision* figuratively, to denote the agent's search space. This could be entirely local, or global, or spread over a network, or confined to an organization. It could be literally ocular, auditory, olfactory, or text based, and so forth. The model permits high heterogeneity of vision. And it would be reasonable to explore this in future research, since people, in fact, differ widely in search spaces, and for a variety of reasons. Some types of information are expensive, for example. Some individuals are simply more inclined to acquire and process information than others. Cacioppo and Petty (1982) dub this "the need for cognition" and present experimental research that could be imported into the agent population. Instead of using a single global value for the sampling radius, one could use an empirically based distribution of vision as a crude analogue of this need for cognition. This could be a nice example of *computational social neuroscience*, where individual agents are based on experimental neuroscience, but then interact with one another in simulated populations.

Run 4. A Day in the Life of Agent_Zero: *How Affect and Probability Can Change on Different Time Scales*

Before we take up the topic of memory in Part III, which also involves time scales, I would like to show how the model can capture three easily recognized spatially explicit examples in which affect and probability change on different time scales. Obviously, many other examples will come readily to mind.

Case 1: Daily Grind

We've all (presumably) had the following experience: we begin the day in a perfectly good mood, go to work, have a lousy day, and come home in a rotten mood. Can we grow this prosaic example? Yes. In Figure 43, Agent 0 starts the day at 7:00 a.m. in his pleasant yellow neighborhood. His affect is zero and his appraisal of the probability of annoying demands is also zero. (Figure 43 also shows the *NetLogo* Interface with its user-adjustable sliders).

Then, as shown in Figure 44, he spends an aggravating and aversive nine-to-five day at the office (located in the upper right quadrant), where he is peppered by annoying demands (orange events). Within hours, his expectation of further annoyance and his aversion (affect) increase until, at quitting time, he is in an absolutely foul humor.

He arrives home, in Figure 45, where he is utterly free of harassment. He knows (since located in the yellow zone) the likelihood of further badgering to be zero, so his P-value drops to zero. But unless extinction is very fast, he is still in a foul humor (high V) when he arrives home. (Maybe his disposition to have a drink even exceeds his threshold!)

Exactly the same run can be interpreted variously.

Case 2: Emergency Responder

For example, one could interpret this as a story about first responders who enter a burning building—a terrifying experience during which the probability of being burned is high in proportion to the frequency of flames

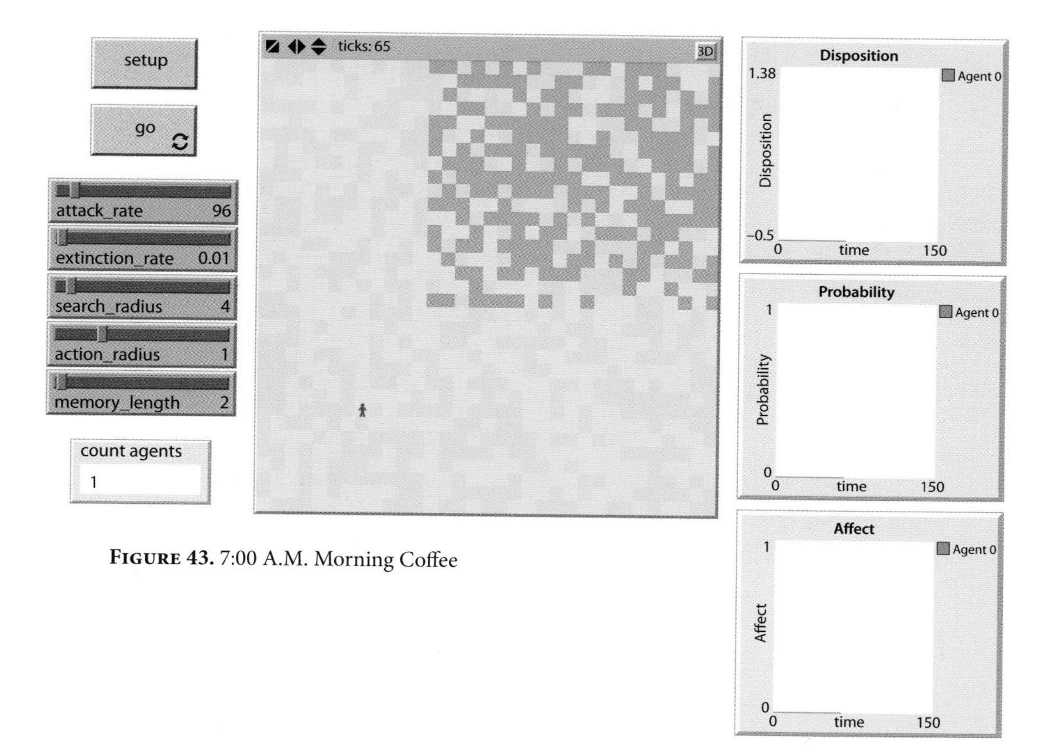

FIGURE 43. 7:00 A.M. Morning Coffee

(orange squares). Once out of the building, the responder knows that the probability of burn injury is zero, but this fact does not extinguish the fear, which endures.

Case 3: Combat

As the most extreme example, one thinks of entry into a war zone with orange bursts as enemy fire. During battle, fear and the probability of being hit are at their maximum. Upon withdrawal from the field, the probability drops to zero, but the posttraumatic stress can endure.

In all three cases, our agent begins the story in the placid yellow zone and in the affectively neutral state: that is, $v(0) = 0$, as shown in Figure 43.

After 150 periods, he ventures to the northeast quadrant—variously interpreted as rife with annoying office demands, flames, or enemy attacks. In this zone, both his affect and his estimate of the probability of aversive events quickly rise to high levels, as shown in Figure 44.

At period 400—quittin' time—our protagonist departs this zone and heads back to home/base. He recognizes that the probability of further

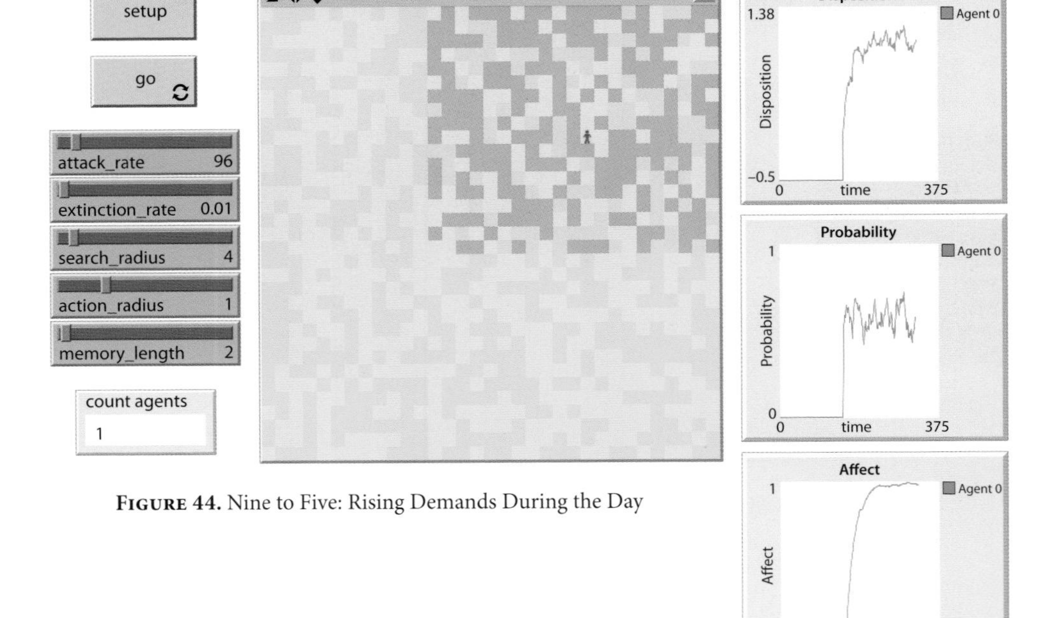

FIGURE 44. Nine to Five: Rising Demands During the Day

adverse events is zero (the sample probability curve falls to zero). But, this is insufficient to reverse his bad feelings, due to a low extinction rate, yielding the results of interest. As shown in Figure 45, Probability drops to zero, but the aversive affect persists.

Of course, as discussed briefly before, by situational conditioning, the agent will come to associate the workplace itself with aggravation. So, *Agent_Zero* is already aggravated (or afraid, depending on the interpretation) when he walks in the door.

Case 4: A Happy Day

Clearly, the preceding fire and combat interpretations would involve fear and the amygdala, among other regions. But the same general associative learning *model*—though not the same brain regions—could apply to happy days, where one's disposition to break out in song is low at the start (Figure 43). So, suppose Agent 0 leaves home in the southwest for her college reunion somewhere in the northeast. On campus, happy singing breaks out all around (the orange outbursts of Figure 44). Agent 0 is rather shy (has a high sing-along threshold) so would never join in, except that her two best (high-weight)

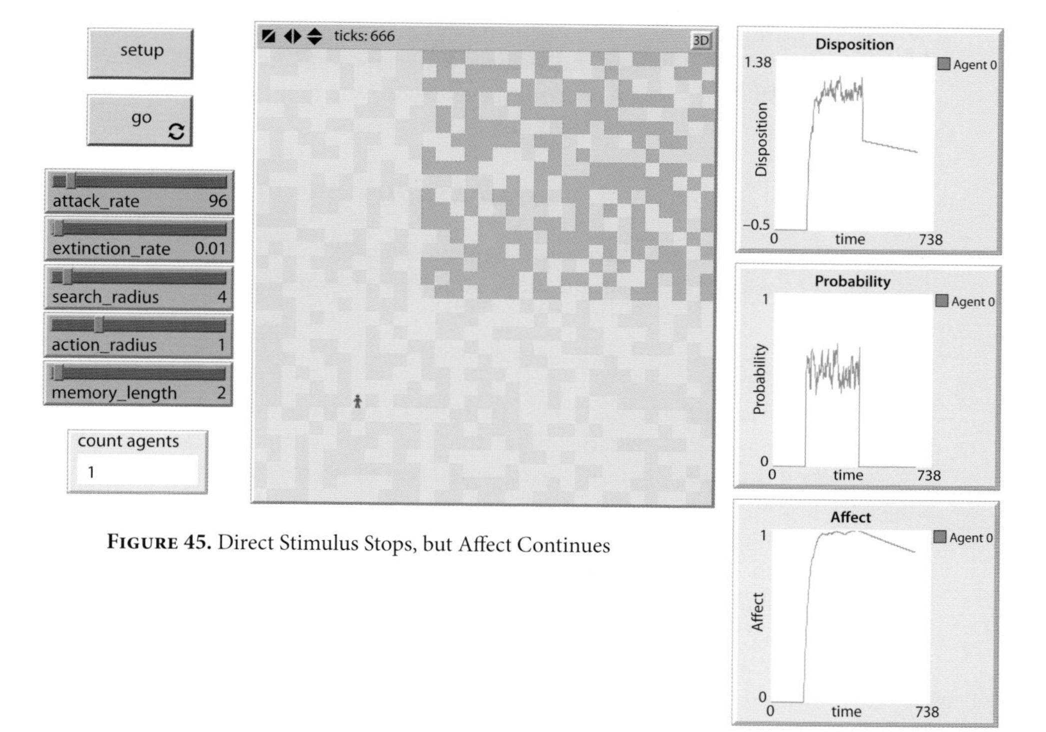

Figure 45. Direct Stimulus Stops, but Affect Continues

college friends (Agents 1 and 2) join in. Their dispositions to sing have weight, so she joins in. Finally, the party ends, and she heads home. And yet, even as the probability of direct musical encounters is zero, she remains aglow and sings the old college songs all the way home, as in Figure 45.[140]

We have been exploring cases where *Agent_Zero's* sample probability rises and falls abruptly with his or her location in space, but her affect persists long after stimuli (trials) end. It would be interesting to devise cases in which affect evaporates before evidence does. We will do so below, when memory—along with much else—is introduced in Part III.

However, perhaps we have shown that the unadorned basic model—the basic *Agent_Zero*—does generate the intended central parables and much more. Specifically, most modeling focuses on extreme events. But the everyday life of people is equally worth modeling and, like the cases we've just developed, can be seen to "ring true" in the model, which is a good start.

Another game one can play with the base model is to explore the effect of one person's deficit on other individuals in her network. Earlier, in

[140] Less frivolously, the model captures cases where the individual has a good impulse but simply needs the support of others to act on it. I thank Julia Chelen for this observation.

connection with posttraumatic stress, we used the mathematical version of the model to explore how one individual's experience affects others. Now, using the agent-based version, we will (I believe for the first time) *lesion* an agent and see the result, not only on her, but on others.

Run 5. Lesion Studies

My limited exposure to the literature suggests the utility of a purely logical dissection of the claims one might make about the amydgala and lesions.

Logic and Lesions

Ever since Klüver and Bucy's (1937) path-breaking work with primates, it had been conjectured that disabling the amygdala virtually eradicates fear. Recalling the rat's apparently hard-wired fear of even cat urine, "Large amygdala lesions dramatically increase the number of contacts a rat will make with a sedated cat. In fact, some of these lesioned animals crawl all over the cat and even nibble its ear, a behavior never shown by the non-lesioned animals" (Davis and Whalen, 2001). More recent lesion studies—or contemporary studies using animals with genetically engineered deficits, such as "knock-out mice"[141]—establish that disabling or eliminating the amygdala indeed eliminates fear (along with much else). So, recognizing many nuances, just for logical precision, let's write this as[142]

$$\neg A \rightarrow \neg F.$$

If no amygdala, then no fear.[143] It follows logically that where there is fear, there is amygdala activation.[144] That is,

$$F \rightarrow A.$$

I have never understood why nature should ever respect our paltry rules of deduction,[145] but this is an observed regularity also (LeDoux 2003). Neither of these entails that excitation of the amygdala causes fear . . . that is, that

[141] See Mayford et al. (1997).

[142] We employ the logic symbols $\neg$ (the negation symbol meaning *not*) and $\rightarrow$ (meaning *implies*).

[143] However, see Cunningham and Brosch (2012).

[144] If p implies q, then not q implies not p. Each implication is the so-called contrapositive of the other, with $\neg A$ playing the role of p.

[145] For example, nature respects Newton's second law, that $F = ma$. But, evidently, it also respects every proposition deducible from this law. But deduction is an entirely human invention. Why should nature select the deducible claims as the ones to which it will physically conform? I find that mysterious.

$$A \rightarrow F.$$

Lesion (or knockout) studies alone show *necessity, not sufficiency,* in other words. However, a history of experiments has shown that, "In humans, electrical stimulation of the amygdala elicits feelings of fear or anxiety as well as autonomic reactions indicative of fear. While other emotional reactions occasionally are produced, the major reaction is one of fear or apprehension." (See Davis and Whalen, 2001, and references cited there.)

While granting, then, that there are experimental grounds for an inference that $A \rightarrow F$, as a general proposition, this is equivalent to

$$\neg F \rightarrow \neg A.$$

which clearly *fails* since fear-inducing stimuli (e.g., snakes) are not the only inputs stimulating the amygdala (A). For example, erotic nude pictures and loud music can activate it[146] (Holland and Gallagher, 1990). So, we are not yet at the point where, given a subject's imagery (even accompanied by many other readings), we can infer their emotional state, or self-reported feeling, if there even is one!

In sum, while the amygdala does exhibit high functional specificity for fear (Kanwisher, 2010), the amygdala is *not* the only brain region involved in fear (Lindquist et al., 2012), *nor* is it the case that the only stimuli that activate the amygdala are fear inducing.

Lesioning *Agent_Zero*

Obviously, lesion studies on healthy humans are unethical. But lesion studies on software people are not (at least not yet). We can knock the amygdala out of *Agent_Zero,* as it were, and explore not only how it affects her behavior, but also how it affects the behavior of all others in her social group!

Here is *Agent_Zero*'s *NetLogo* amygdala, speaking very figuratively:[147]

```
[
if pcolor = orange + 1
  [set affect affect + (learning_rate * (affect ^ delta) * (lambda − affect))]
if pcolor != orange + 1
  [set affect affect + (learning_rate * (affect ^ delta) * extinction_rate * (0 − affect))]
]
```

[146] This indicates that the amygdala is, in fact, not specialized to fear. It is implicated in many kinds of arousal.

[147] Again, I am not modeling brain regions.

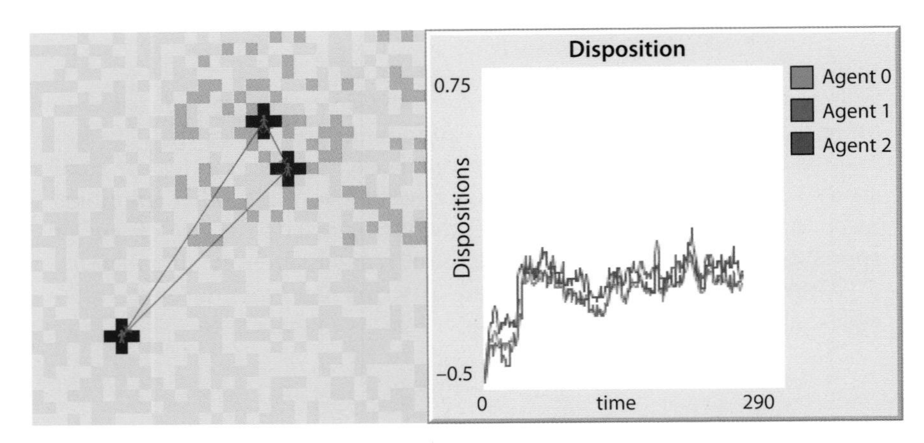

FIGURE 46. All Agents Normal

It is the agent's update-affect routine. In English, it says, "If the patch you're on bears the adverse event color, then set your new affect to your old affect plus the product of: (a) your learning rate, (b) your old affect to the delta power, and (c) the difference between lambda and your old affect. Otherwise (i.e., if the patch does *not* bear the adverse event color), do as before but replace (c) with the extinction rate times the negative of old affect, all of which follows the Rescorla-Wagner scheme.[148]

To knock out an agent's amygdala, we simply knock out this *NetLogo* Code block.[149] We are interested not only in how this lesioned agent behaves, but also in how her neurocognitive deficit affects the whole network. Depending on the agent's weight, the effect can be dramatic. One agent's deficit can have far-reaching ramifications. In Figure 46, we see a Run with all agents functioning normally.[150]

If we now knock out the amygdala of Agent 2 (the upper-right rover), it eliminates her fear (and her violence) and her transmission of fear to both Agents 0 and 1. Agent 1 (upper left) still acquires fear directly from events and transmits this to Agent 0. But lesioned Agent 2 is no longer contributing to Agent 0's fear (either directly or through

[148] My code actually generalizes Rescorla-Wagner extinction slightly by introducing the variable named extinction_rate, which is a user-adjustable slider in the *NetLogo* Interface. If extinction_rate = 1, then the scheme is exactly the classical Rescorla-Wagner model. If $0 < \text{extinction\_rate} < 1$, slower extinction trajectories can be explored. Typically, $\text{extinction\_rate} \in [0, 1]$.

[149] This is the sense in which *Agent_Zero*, as a software object, is "modular," which is to make no claim whatever regarding the modularity (however defined) of the human brain.

[150] The slider settings in this case are: Attack Rate 25, Extinction Rate 0, Sampling Radius 4, Action Radius 1, Memory 1, with Seed 2, which are also given in the Table of Appendix IV.

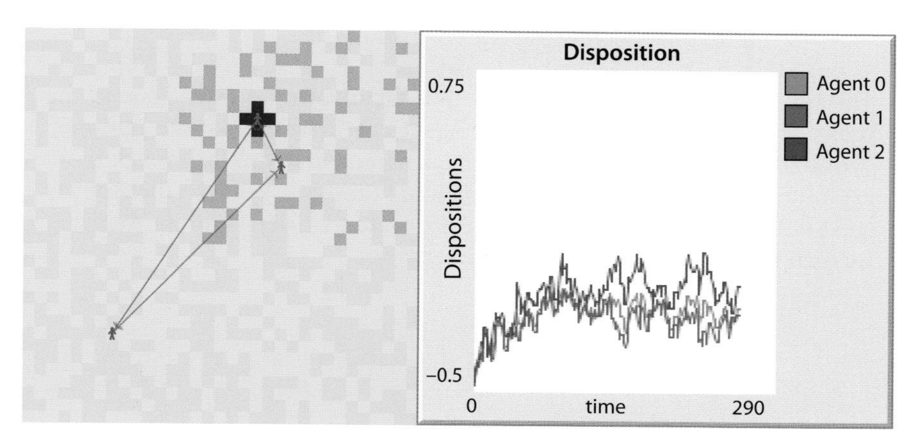

FIGURE 47. Agent 2 (Upper-right) Lesioned

Agent 1). As a result, the total fear acquired by Agent 0 is now beneath her action threshold, and she never engages in violence. This is shown in Figure 47.

Emotional contagion dynamics are affected if one agent's *direct* fear acquisition is disabled. But, as we now discuss, *observational* fear acquisition may also be impaired by amygdala damage.

Patient S. M.

The famous subject S. M. suffered from Urbach-Wiethe disease. In their classic paper, Adolphs et al. (1994, p. 670) write that her condition "caused an nearly complete bilateral destruction of the amygdala, while sparing hippocampus and all neocortical structures, as revealed by detailed neuroanatomical analyses of her computed tomography (CT) and magnetic resonance imaging (MRI) scans." The result was that S. M. was unable to recognize fear—and emotion generally—in the faces of others. To represent S. M. in the *Agent_Zero* framework, we would add to the disability just discussed the further inability to acquire fear *observationally* (as discussed earlier). Mathematically, this would be arranged by zeroing out the weight S. M. assigns to the affect of others.[151] In the social setting, this will further damp network transmission because she will not pick up emotion; and so she will not pass it on either. So, the damping social effect would be even more pronounced. That, at any rate, would be the hypothesis.

[151] Technically, weight has thus far been assigned only to the sum of V and P.

Generative Minimalism

The runs and discussions presented thus far involve no extensions to the basic *Agent_Zero* model. While the agent specification is quite minimal, considerable generative capacity has been demonstrated. While much more exploration of the basic model is warranted (and is easy given the Applets and Source Code posted on the book's Princeton University Press Website), we turn now to 14 significant extensions.

PART III

Extensions

The 14 extensions developed here are as follows:

1. Endogenous destructive radius
2. Age and impulse control
3. Flight rather than fight
4. Replication of the Latané-Darley experiment
5. Introduction of memory
6. Coupling of affect and cognition
7. Endogenous changes in weight strength through affective homophily
8. Growing the Arab Spring
9. Jury processes
10. Emergent dynamics of network structure
11. Vertical structure and multiple scales
12. *The 18th Brumaire of Agent_Zero*
13. Introduction of prices and seasonal economic cycles
14. Endogenous mutual escalation spirals

III.1. ENDOGENOUS DESTRUCTIVE RADIUS

Just as agents can differ in their search radii, so they may differ in their destructive radii. Thus far, this has been treated as a single exogenous global constant. It is more realistic—and reduces the number of freely adjustable parameters—to endogenize this action radius. It might, for example, be a function of affect, or of total disposition.[152] In Figure 48, the destructive

[152]The coding change required to endogenize the destructive radius as a function of positive affect is trivial:

```
to take-action
  ask turtles [
    if disposition > 0 [ask patches in-radius (affect * 10) [set pcolor red – 3]]
    ]
end
```

Here, 10 is just a scaling factor and destruction is color-coded as pcolor red-3.

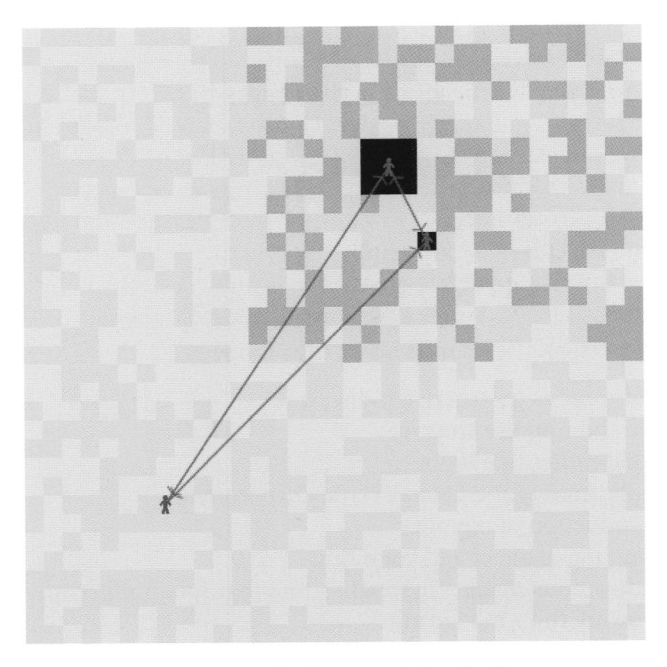

FIGURE 48. Endogenous Action Radii

radius is a simple linear function of disposition and thus differs among agents (and varies in time). This is a fertile extension, especially in the various alternative interpretations of the framework.

Health Interpretations

For example, where the action is vaccine refusal—itself an area rife with emotional contagion[153]—some agents foreswear all vaccines within a large pharmaceutical radius, while others refuse only a narrow set (a small radius). In an eating interpretation, some indulge in bingeing across many food groups, while others confine themselves to one. In the obesity interpretation, for example, a large binge radius may be the result of a specific deficiency in dopamine receptor availability.[154]

[153] For a well-researched popular account, see Mnookin (2011).

[154] Regression of dopamine receptor density on BMI exhibits a negative linear relation (Wang et al., 2001).

Economic Interpretation

Now interpret yellow patches as healthy financial assets, and neighboring yellow patches as "similar" ones (stocks in the same industry, for example). Assets turn orange when they suddenly lose value. The destructive radius is the set of assets dumped in response. This can certainly grow with fear and contribute to cascading crises through contagion. In Figure 48, the right-most agent's affect is quite low and his destructive radius is confined to one orange site. But the upper left agent's fear is high, so in response to a single orange devaluation, he dumps a large radius, including healthy yellow assets.

III.2. AGE AND IMPULSE CONTROL

Large action radii might well be interpreted as evidence of poor impulse control or some specific deficiency in executive function. For example, in distinguishing between minors and adults, the U.S. legal system recognizes a difference in juvenile and adult impulse control. The psychology literature surrounding impulse control over the life course is large but reinforces the general pattern: the younger you are, the worse your impulse control (Mischel et al., 2011). Thus far, agents have not advanced in age. So, here we'll do a double extension: we'll have them age, but we'll also have impulse control increase with age. We will think of impulse control as the gap between one's affect and one's destruction. For agents with poor control, the damage is far out of proportion to affect, while mature agents can align the two. Specifically, let us posit that when one is born (age zero), impulse control is nil, and one's damage radius exceeds one's affect by a factor $k > 0$. If this impulsive excess is assumed to shrink linearly with age, we arrive at the following functional form:

$$\text{Damage\_Radius}(t) = \text{Affect}(t) + k\left[1 - \frac{\text{Age}(t)}{\text{Maximum\_Age}}\right]. \qquad [38]$$

NetLogo Code for aging and for this function are provided in the relevant Applet. But, the younger you are, the more your damage exceeds your affect (i.e., the worse your impulse control). The two approach equality as agents approach the maximum age. Figure 49 gives a simple example. All three agents are stationary in the active quadrant, and so are getting comparable stimuli. The upper agent goes first at age 8. The lower agent goes second at age 22, and the third goes at age 36. Their radii are successively smaller, reflecting their increased impulse control with age.[155]

[155] To generate this particular picture, use the Applet provided on the Princeton University Press Website, with a random seed of 2, as per Appendix IV.

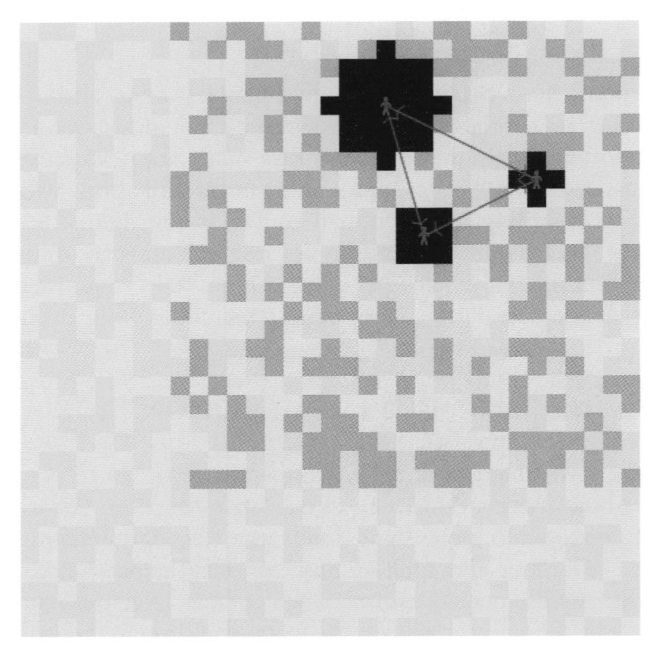

FIGURE 49. Age and Impulse Control

Obviously, a wide variety of other effects of age can be modeled. Indeed, impulse control itself might be unimodal, declining again with age (e.g., with dementia). Beyond cognitive effects, loss of visual acuity (e.g., contrast sensitivity)[156] could be directly represented as a reduction in the spatial sampling radius, while decline in aerobic capacity[157] could be represented as a reduction in the distance traveled per step. The general loss of mobility with age obviously affects the elderly's ability to evacuate in disasters or migrate from politically or environmentally hostile areas, which brings us to the topic of flight in general.

III.3. FIGHT VS. FLIGHT

In the exposition thus far, agents have responded to aversive stimuli with direct action on a neighborhood of sites. Destruction—a "fight" response—has been our prime example. Another possible response, of course, is

[156] Haegerstrom-Portnoy, Schneck, and Brabyn (1999).

[157] Harms, Cooper, and Tanaka (2011).

flight. A reasonably general model should permit both fight and flight. The latter is easily arranged in our framework. In fact, the *NetLogo* Applet provided on the Princeton University Press Website offers users a simple switch. If it's in the "on" position, agents fight (destroying aversive patches). If it's in the "off" position, they flee aversive stimuli. A direct comparison of fight and flight cases—where everything else is the same—is quite interesting.

Case 1: Fight

Beginning with the fight case, for illustrative purposes, we will increase the retaliatory damage radius to 2 in each N, S, E, and W direction so that 12 sites are taken out. With agents in fixed positions (since we are contrasting to flight), a representative run is depicted in Figure 50.

Movie 5 shows that for the arbitrary settings (see the Parameter Table, Appendix IV) employed, the agents fight (retaliate on yellow sites) in a certain order: the southern agent goes first, then the western, and then the eastern. The last is moved to do so through the weighted (solo) dispositions of the first two.

Figure 50. Fight [Movie 5]

Case 2: Flight

This is also the same sequence in which the agents begin to flee, as shown in the four snapshots of Figure 51.

I highly recommend the movie version, **Movie 6** on the book's Princeton University Press Website, because it clearly shows the upper-right agent being "dragged" out by the other two. This is the analogue of them "convincing" him, through their weights, to finally act as they have.

So, here are two runs in which the only change is flight versus fight. Everything else is identical, including the random seed and all the stochastic

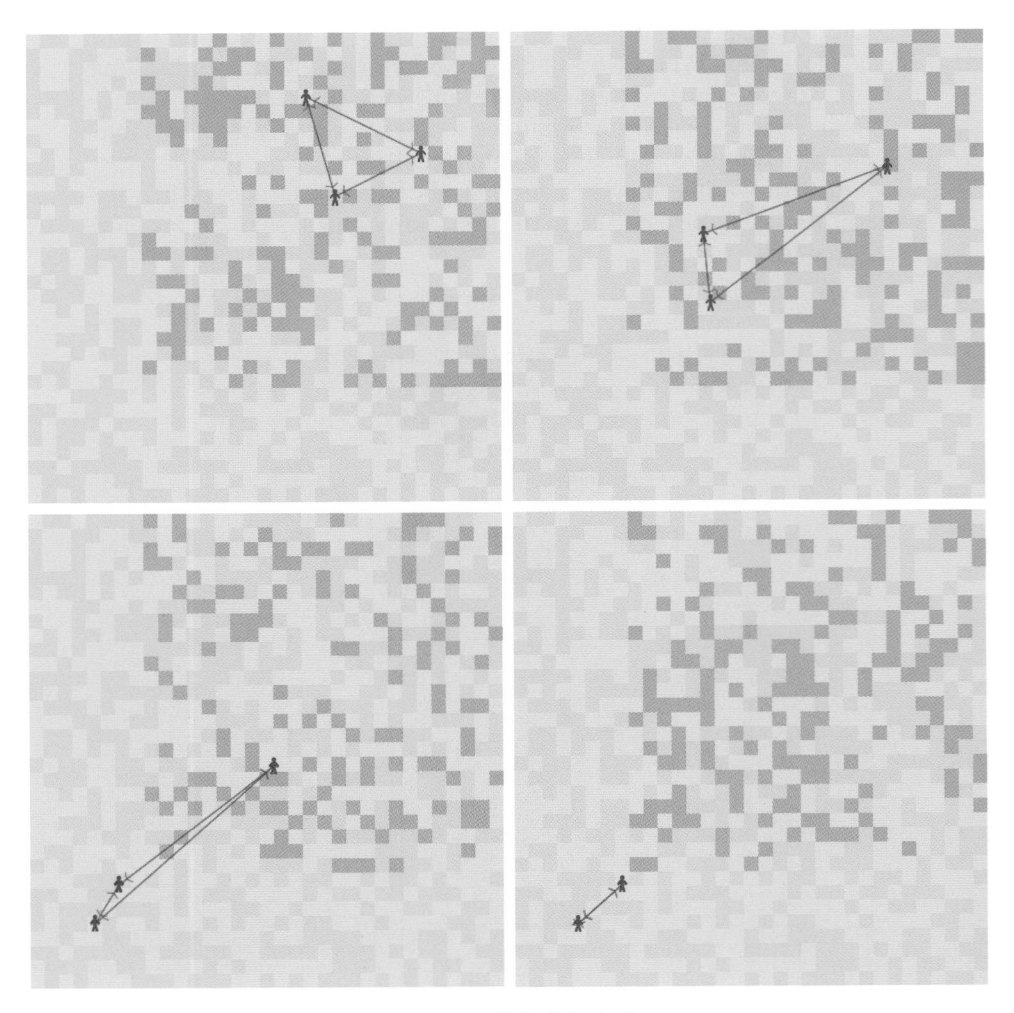

FIGURE 51. Flight [Movie 6]

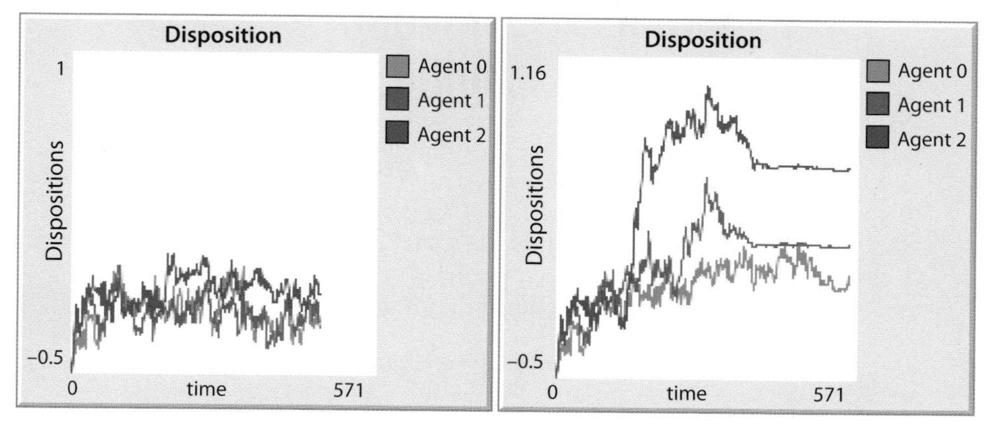

FIGURE 52. Fight vs. Flight Dispositions Compared

activations generated. Suppose we now ask the question: How, if at all, do the disposition trajectories differ in these two (fight vs. flight) runs? And, if they differ, why?

They differ radically, as shown in Figure 52. Under fight (left), net disposition exceeds zero briefly but then remains low throughout. Under flight, it grows dramatically, settling down to sustained levels only when agents are clear of the entire stimulus area. Why are they so different? Because the stationary fighters eliminate the stimulus and never move into new active areas. The refugees, by contrast, cross the entire field of active stimuli as they evacuate, as it were, continuously increasing their affect, their probability, and the weighted sum of their colleagues' values, all of which is aggregated to form their disposition. If evacuation subjects the refugee to a relentless field of adverse stimuli, shelter-in-place may be the less traumatic course—if it is feasible.

An area where the dilemma of shelter-in-place vs. evacuate arises very sharply is in chemical, biological, and radiological contamination scenarios (Fischhoff, 2005). Obviously, evacuation through a contaminated zone might be riskier than staying put. But if the contaminant is purely airborne, then depending on the wind field, the fluid dynamics, and available transport capacity, evacuation might dominate. For an urban-scale simulation combining computational fluid dynamics and agent-based modeling, see J. M. Epstein, Pankajakshan, and Hammond (2011). Mainly, it is clear that the areas of refugee behavior and crisis evacuation could be studied in this framework.

We will return to the topic of flight when we replicate the famous Latané-Darley experiment. There, the orange patches represent smoke, and—unlike the preceding example—once a patch turns orange, it stays orange; the smoke does not dissipate but eventually fills the room.

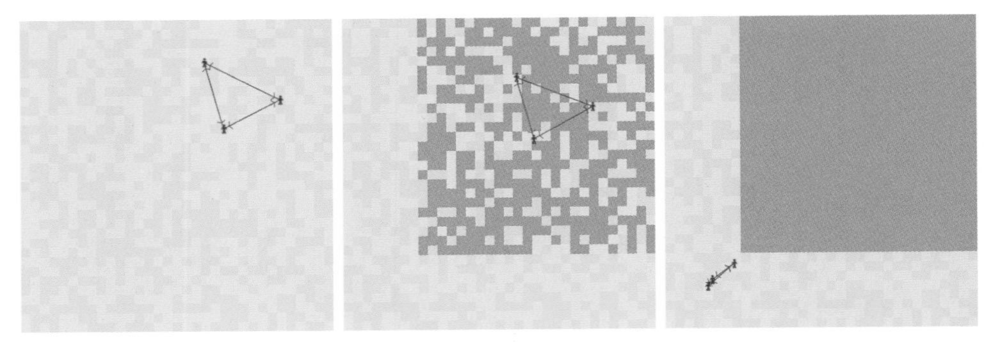

FIGURE 53. Capital Flight

Capital Flight

Such situations also occur in economics. Some assets (e.g., houses) can lose value and never recover. So, imagine agents as investors, each of whose "portfolios" is the set of all patches within their financial vision. Yellow patches have high value. If a patch turns orange, it suddenly loses value—and does so permanently. As the market collapses, investors migrate their portfolios from low-value patches to high-value ones. We see capital flight, in other words, as agents (located at the center of their portfolio) migrate in the southwesterly direction in asset space. Interagent weights capture contagion effects, and the most financially fretful can, just as before, "pull" other agents into financial flight. See Figure 53.

With no recovery of value (no reversion to yellow), the agents encounter even more aversive stimuli along their path. Consequently, everyone's affective, probability, and disposition curves—all amplified by contagion—are even higher than before. Net disposition to flee one's portfolio, shown in Figure 54, contrasts with the preceding two examples.

Physical flight figures centrally in one of Latané and Darley's classic experiments in social psychology. A small extension of the basic model will permit us to replicate this in the *Agent_Zero* framework.

III.4. REPLICATING THE LATANÉ-DARLEY EXPERIMENT

Much empirical work with agent-based models (including my own) aims to replicate large systems, such as infectious disease dynamics, stock-market dynamics, or distributions of firm sizes, city sizes, and the like.

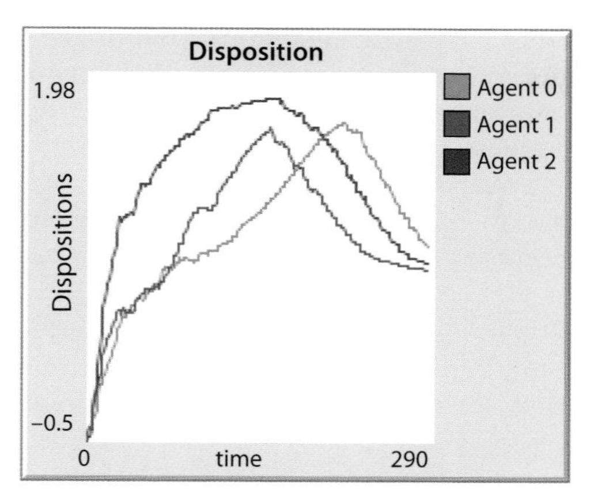

FIGURE 54. Disposition to Flee Portfolio

So, it is natural to question whether a three-agent model could hold the slightest empirical interest. It happens that many of the most famous experiments in social psychology involve only three people. Two of these are the Milgram (1963) experiment and Latané and Darley's 1968 experiment. We will "replicate" the latter of these. That is to say, we will see if *Agent_Zero* can behave essentially as humans behaved in this experimental setting.

The Latané-Darley experiment is very clear. It compares behavior in the absence of others to behavior in the presence of others. In the first case, the subject is seated alone in a room. Smoke begins to enter the room. He becomes agitated, eventually afraid, and when his affect and risk appraisal exceed some level, he exits the room and reports the smoke. In the second preparation, the subject is seated in the room with two strangers (who, unbeknownst to the subject, are confederates of the experimenter). Smoke enters the room, just as before. The confederates take no notice whatever and continue filling out forms. In this case the subject takes much longer to exit the room and report. So, the same physical environment stimulates different behavior. Can we generate that in our simple agent model? With one very simple extension, we can.

Threshold Imputation

Without loss of generality, the skeletal equation for two agents posits that Agent 1's net disposition (subtracting the threshold) is given by

$$D_1^{\text{net}} = V_1 + P_1 + \omega_{21}(V_2 + P_2) - \tau_1. \quad [39]$$

Since Agent 2 is a confederate and knows the smoke to be a ruse, his V and P are both zero and, in the model, should have no effect on Agent 1. Agent 2's own threshold does not figure in Agent 1's disposition, as has been assumed throughout. We appear to be at a loss for any mechanism that would alter Agent 1's behavior. Suppose, however, that Agent 1 were to impute some threshold to Agent 2 and incorporate it in equation [39]. Denote this imputed threshold as ψ_{12}.[158] Then the D_1^{net} equation becomes

$$D_1^{net} = V_1 + P_1 + \omega_{21}(V_2 + P_2 - \psi_{12}) - \tau_1. \qquad [40]$$

With three agents, if we assume $\psi_{12}, \psi_{13} > 0$, we generate the qualitative result of interest.

Using $\psi_{12} = \psi_{13} = 1.5$, Figure 55 offers two snapshots of the experiment recorded by hidden ceiling cameras (as it were) looking down on events. The orange patches are smoke filling the room.

The solo agent of Figure 55 exits after around 25 time intervals, while in the trio treatment (right panel), he takes three times as long. These screen shots were taken just as the subject exits the room. Notice that there is less smoke on the left than on the right. Complete movies of the solo and trio cases are given as **Movies 7 and 8** on the book's Princeton University Press Website.[159]

So, in a very simple sense, we have replicated Darley and Latané's basic qualitative result: Subjects behave differently in the two cases, and we have the same direction of change as they observed.

Now, *why* is this happening? Since they know the smoke to be a ruse, the confederates' affect (their Vs) and their appraisals of risk (their Ps) are both zero. So, how are they influencing the subject's behavior at all? What mechanism, in the *model* (as against the brain), is producing the different behavior in the two cases? Threshold imputation is the mechanism.

Without loss of generality, assume Agent 1 to be the subject. For him to flee the room when alone, all we require (suppressing time) is that $V_1 + P_1 > \tau_1$. But imputing thresholds to others, his disposition becomes

$$D_1^{net} = V_1 + P_1 + \omega_{21}(V_2 + P_2 - \psi_{12}) + \omega_{31}(V_3 + P_3 - \psi_{13}) - \tau_1. \qquad [41]$$

For the confederates, $V_2 = P_2 = V_3 = P_3 = 0$, leaving

[158] The subscript order is reversed to suggest that, unlike the experience of weight, Agent 1 is actively imputing a value.

[159] Of course, each run of the model is a single sample path of a stochastic process. And we should never assume a particular realization to be statistically representative. But, in this case, we will prove analytically that the central *ordering* holds up for positive threshold imputations.

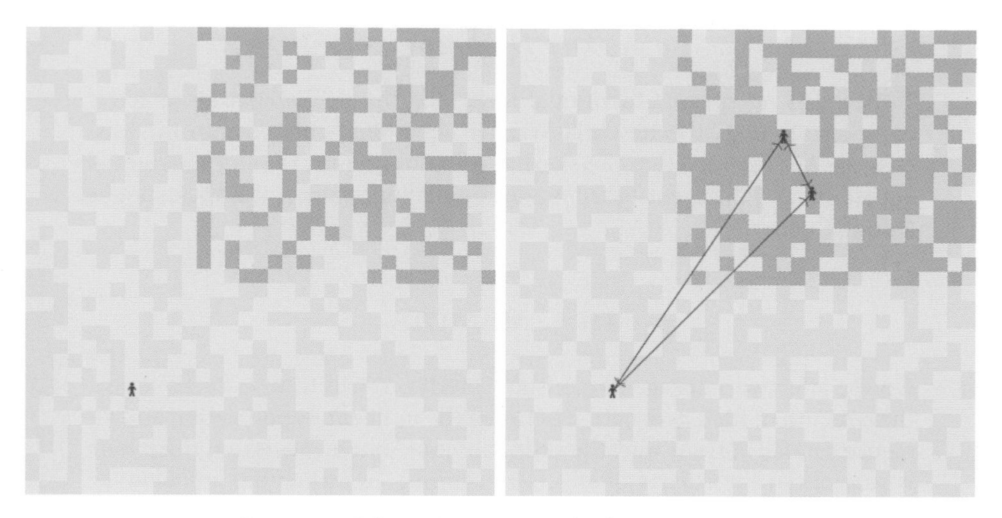

FIGURE 55. Solo vs. Trio Latané-Darley [Movies 7 and 8]

$$D_1^{\text{net}} = V_1 + P_1 - \omega_{21}\psi_{12} - \omega_{31}\psi_{13} - \tau_1. \qquad [42]$$

For D_1^{net} to be positive in this case, we require not merely that $V_1 + P_1 > \tau_1$ but that $V_1 + P_1 > \tau_1 + \omega_{21}\psi_{12} + \omega_{31}\psi_{13}$. This is a *higher threshold* as long as ψ_{12} and ψ_{13} are positive and, accordingly, it takes longer to exceed it![160] This is plausible. First, simply as fellow humans, they have some, perhaps tiny, weight on the subject agent (indeed it suffices for only one of their weights to be positive). Second, even though they are confederates, there is some level of danger that would stimulate them to act, or at least, so thinks any normal Agent 1. So, as long as he imputes even the smallest threshold to them, his overall threshold will rise in their presence. In fact, he will more likely impute high thresholds to them, making for the dramatic difference we see! Here, then, is a simple generative mechanism for bystander effects.

Notice that threshold imputation retrodicts that a single confederate, rather than Darley and Litane's original two, can suffice to delay the subject's exit,[161] which they also found (Latané and Darley, 1968). Ceteris paribus, we would further retrodict that the effect would be weaker in the single confederate case—the subject should leave later than when alone, but earlier than when there are two confederates. This, too, was found.

[160] Assuming monotonicity in time, of course—a weak requirement.

[161] This follows from equation [40]. With $V_2 = P_2 = 0$, the requirement becomes $V_1 + P_1 > \omega_{21}\psi_{12} + \tau_1$, which exceeds τ_1.

The Dialogue

Notice that no specific functional forms for the deliberative or affective components (e.g., the Rescorla-Wagner equations) entered into this discussion. We thus see that the skeletal equations alone, the full differential equations, and the agent-based model are all illuminating, as is the dialogue among them.

III.5. MEMORY

On the affective side, agents can show inertia. When trials (attacks) cease, the strength of their association between the indigenous population and violence can fall at some rate, the extinction rate. If this rate is low, then there is substantial persistence of affect. By contrast, in estimating *probabilities*, the agents developed thus far have no memory, or inertia, whatsoever. At every iteration, they use only the current relative frequency of orange (to total) agents within their "vision" to estimate the attack probability.

This seems somewhat implausible for humans. If the attack (i.e., adverse event) rate has been consistently high for the last 10 days, the mere fact that it is zero today may not lead us to forget the entire history and assume a rate of zero for tomorrow. We certainly might *reduce* the expectation of attack, but the history casts some shadow on projections of the immediate future. Exactly how it does so is a vastly complex matter and depends on the history and the type of memory.

Since I am conceptualizing this as part of *Agent_Zero*'s deliberative, empirical, data-based, component, it would fall under what neuroscientists call *declarative memory* capacity. And since it involves a sequence of discrete aversive occurrences, it would engage what is termed *episodic* memory.[162] Episodic memory undoubtedly involves a widely distributed network of brain systems. But the hippocampus and parahippocampus are centrally implicated in this capacity. The latter is hypothesized to play the greater role in the immediate updating of the event history, which is stored more generally in working memory controlled by hippocampus (Eichenbaum, 2000; Smith and DeCoster, 2000). On neural mechanisms of memory, see Kandel (2001).

As emphasized throughout, I am not modeling these (or any other) brain regions. My aim is to offer a simple plausible model of performance, shaped by our evolving knowledge of process and by the broader empirical literature. For these purposes, many mathematical—signal processing—strategies

[162]See Smith and DeCoster (2000, p. 108): "Humans possess 2 memory systems. One system slowly learns general regularities, whereas the other can quickly form representations of unique or novel events."

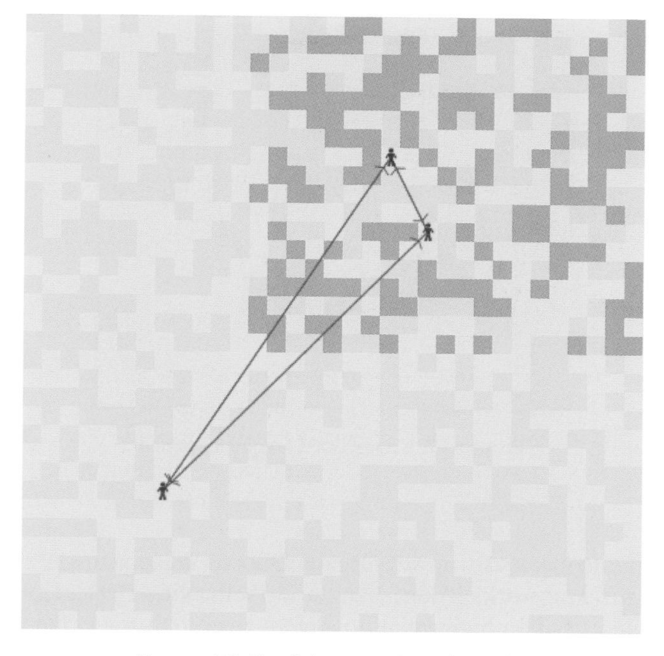

Figure 56. Fixed Agents in Stimulus Field

present themselves. In selecting one, let us begin by recalling the kind of signal *Agent_Zero* gets from his stochastically aversive environment over, for instance, 100 periods.

Figure 56 gives a typical snapshot of the agents. Agents 1 and 2 are in the hostile northeast quadrant of our space. The attack rate is 50.

Each period they "calculate" the relative frequency of orange outbursts within their vision (here vision = 4). Their resulting frequency time series are shown in Figure 57.

There are two things to notice, one individual, and one social. First, the signals for Blue and for Red are highly variable.[163] And, depending on action thresholds, this might produce equally variable action dispositions and wildly erratic behavior, whipsawed by environmental variability. This happens, but one's model should at least *permit* the representation of less-jittery behavior. The second thing to notice is social: with memory zero in a spatially stochastic environment, the (spatially proximate) agents, in addition to being volatile, are only weakly correlated. Presented with signals of the sort plotted in Figure 57, the range of methods for extracting underlying regularities is vast, from Fourier analysis to extract periodicities, as

[163]The agents are motionless and their spatial sampling radius is 4.

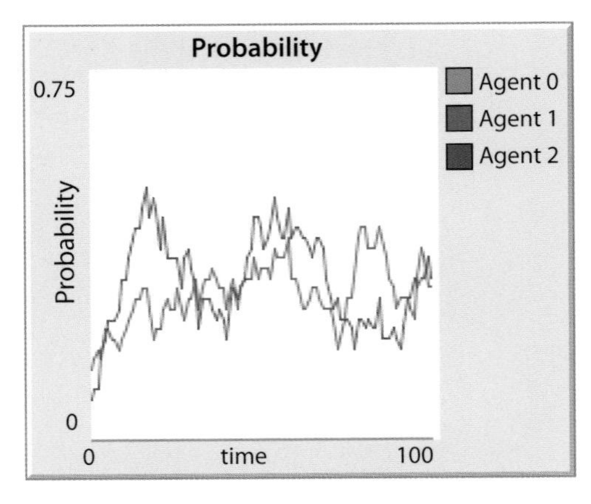

Figure 57. Unprocessed Frequency Signal

conducted unconsciously by our Organ of Corti (Wagenaar, 1996) to wavelets and neural networks (Akay, Akay, and Welkowitz, 1994). However, the simplest filter used on high-volatility data is the moving average. Here, one uses the average of the most recent *m* readings, where *m* is the memory, also called the sampling window or reading frame. So, if $m = 10$, one uses the most recent ten observations. Every time step (e.g., day), the most recent is added, and the least recent is dropped.[164]

Before, the agent used simply his estimate of relative frequency within the spatial sampling radius (v) at each time, $RF_v(t)$. Using the moving average over the preceding *m* periods, his probability estimate becomes

$$P(t) = \frac{1}{m}\sum_{t-m}^{t} RF_v(t), \qquad [43]$$

which obviously reduces to the previous case for $m = 1$ (i.e., processing only the current period). This produces smoother dynamics for each individual and (for our neighbors) greater consensus among them, as shown in Figure 58, for a memory window, or "reading frame," of 25 periods (all other parameters are as in Figure 57).

Neighboring agents standing in a turbulent bathtub will experience very different signals on short time scales. But over a large window, their average signals will be more closely correlated.

[164]This is used, for example, in the game theoretic model of Best Reply to Recent Sample evidence (Axtell, Epstein, and Young, 1999, pp. 191–211).

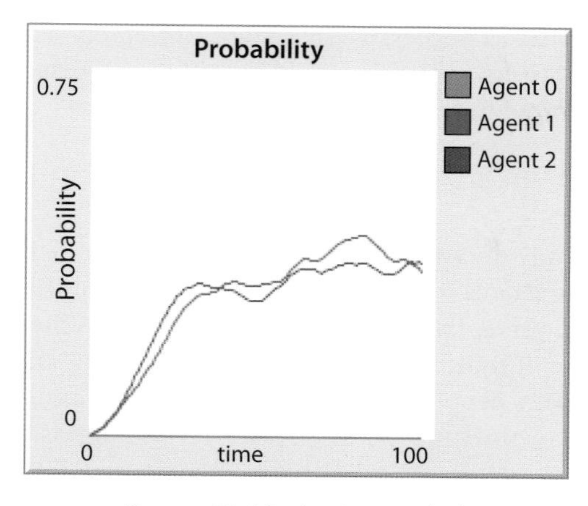

FIGURE 58. Moving Average (25)

Limits of the Moving Average

Numerous refinements are possible. For one, in a moving average each memory receives equal weight, when memory can decay with time, an effect often handled with linear or exponential weighting.[165] More interestingly cognitively, there can also be powerful anchoring effects (Tversky and Kahneman, 1974). My favorite of their demonstrations works as follows: Different subjects are read a string of numbers and asked to estimate their product. When a series of integers is presented from highest to lowest, their product is systematically estimated as far larger than when the same sequence is presented from lowest to highest. That is, if numbers are read aloud in the order $8 \times 7 \times 6 \times 5 \times 4 \times 3 \times 2 \times 1$, the product is systematically estimated as being far larger than if the numbers are announced as $1 \times 2 \times 3 \times 4 \times 5 \times 6 \times 7 \times 8$. We "anchor" on the first number we hear—the 8 vs. the 1 (Tversky and Kahneman, 1974, p. 1128). The moving average is insensitive to presentation order. One could capture anchoring by assigning a maximum weight to the first item presented and then having weights

[165]Technically, a moving average (MA) can also show decay. Suppose that we remember the most recent six events and that we receive a string of exactly six values, each of which is a 1. Assume that, from then on, we get no new input. How does the moving average evolve? The opening memory array would be (1, 1, 1, 1, 1, 1), and its MA is $\frac{6}{6} = 1$. Then the array is (0, 1, 1, 1, 1, 1) and the MA is $\frac{5}{6}$. Next period, memory holds (0, 0, 1, 1, 1, 1), and MA $= \frac{4}{6}$. Then the values will be $\frac{3}{6}, \frac{2}{6}, \frac{1}{6}$, and, finally, 0. So, it does decay. For a point event, it simply drops to zero, and some form of weighting, such as exponential, could be introduced. It would then be interesting to explore cases where this exponential decay and the exponential decay of affect differ in various ways or are equal, a suggestion for which I thank Robert Axelrod. Obviously, I am not claiming to have resolved this issue, only to show that the *Agent_Zero* framework includes memory, and can accommodate a wide range of treatments.

successively decline as successive items are presented. The recency effect (e.g., Deese and Kaufman, 1957) would be just the reverse, with maximum weight always shifted to the most recent item presented.

Moving Median

Another simple way to confer stability on the agent's remembered distribution of sample estimates is to use the moving median rather than the moving average. Of course, the moving median ignores outliers—such as new extreme evidence. It is thus less vulnerable to anchoring. In some cases, this is realistic, as when people fanatically stick to their opinions despite new counterevidence. Sometimes, however, we do the opposite, giving undue weight to extreme events. And, of course, people may differ as well in the length of their memory and in their processing of it.

The *NetLogo* Code allows the user to choose any memory window and either the moving average or moving median. We will use the moving average in a number of extensions that follow, perhaps most colorfully in extension XII, *The 18th Brumaire of Agent_Zero*. But we turn now to a set of couplings *between* components of *Agent_Zero*.

III.6. COUPLINGS: ENTANGLEMENT OF PASSION AND REASON

Thus far, I have modeled the affective, deliberative, and social components as independent; they all affect disposition, but none depends on—is an explicit mathematical function of—the others. They are decoupled. But, in fact, we have considerable evidence that they are entangled. Here I explore some simple ways of modeling the influence of affective dynamics on (a) probability estimates and (b) network structures. These extensions, neither of which increases the number of freely adjustable parameters in *Agent_Zero*, are as follows:

1. Probability estimate is influenced by affect.
2. Interagent network weights vary with affective strength and homophily.

Emotional Amplification of Probability Estimates

Throughout, I have modeled *Agent_Zero*'s probability estimates as being independent of his own affect. To be sure, the probability estimates have been biased. But it has been a sample selection bias. Affect, per se, has not been a source of bias. One of the more interesting and well-established things about humans is that our emotions *do* bias our estimates of relative frequency—and do so in a systematic direction. In general, this phenomenon falls under what Paul Slovic termed the *affect heuristic* (Slovic et al., 2007; Kahneman, 2011). More recently, Lerner, Gonzales, Small, and Fischhoff (2003) found that "In a nationally representative sample of Americans ($N = 973$, ages 13–88), fear increased [terrorist] risk estimates. . . ."

So, what is the simplest way to extend the model to permit this? Mathematically, we require that the higher the emotion (e.g., fear), the greater the upward bias. However, if the emotionally neutral probability estimate is already 1.0, there is no possible increase, whereas if the initial estimate is close to zero, the room for upward bias is great. If affect is at its absolute maximum, let's assume that probability is estimated at 1.0, although lower upper bounds are plausible.

In addition to these functional desiderata, let's also avoid introducing any new variable into the model. One simple way to do all this is as follows: If P_n is the emotionally neutral estimate (which may still be biased statistically), then the emotionally affected estimate, P_e is given by

$$P_e = P_n^{1-V}. \qquad [44]$$

When V is zero, there is no emotional bias and $P_e = P_n$. When V is 1, there is maximum emotional bias and $P_e = 1$.

This works nicely, as the plots in Figure 59 suggest. The y-intercept is the initial estimate of P and always ranges from 0 to 1; we show different plots for different P_n values. The greatest impact of emotional bias occurs where the P_n is lowest. This accords with the experimental results of DeSteno, Petty, Wegener, and Rucker (2000).

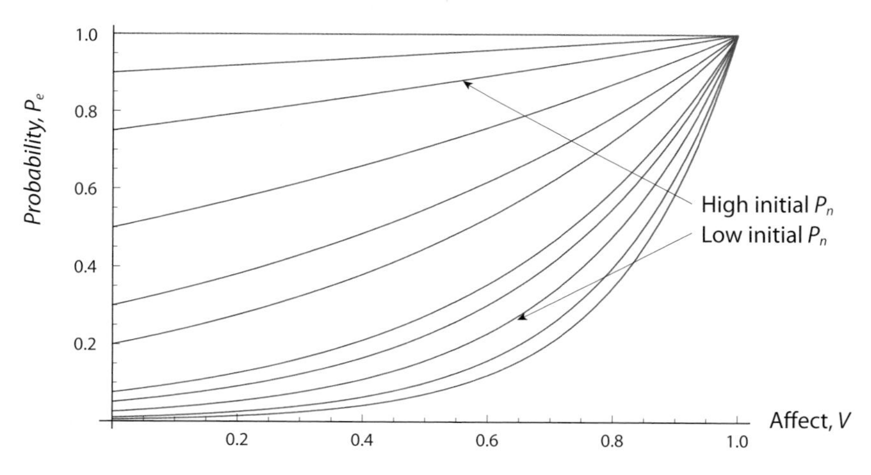

FIGURE 59. Probability Estimate as Function of Affect

Mathematical Treatment

We will introduce this variation into the agent model shortly. But, just to keep the agent-based model honest, as it were, we will first incorporate it into the mathematical version. Recall the skeletal equation[166]

$$\forall i, D_i^{\text{tot}}(t) = V_i(t) + P_i(t) + \sum_{j \neq i} \omega_{ji}(V_j(t) + P_j(t)). \qquad [45]$$

Here, affect (V) has no effect on probability judgment (P). If we now introduce the preceding affect-dependent variation, we obtain

$$\forall i, D_i^{\text{tot}}(t) = V_i(t) + P_i(t)^{(1-V_i(t))} + \sum_{j \neq i} \omega_{ji}(V_j(t) + P_j(t)^{(1-V_j(t))}). \qquad [46]$$

Now we can ask, What is the effect of Agent j's affect on Agent i's total disposition? This is the cross-partial derivative, $\partial D_i / \partial v_j$.

$$\frac{\partial D_i}{\partial V_j} = \omega_{ji}\left[1 + \frac{d}{dV_j}(P_j^{(1-V_j)})\right] = \omega_{ji}[1 - P_j^{(1-V_j)} \ln P_j]. \qquad [47]$$

But the limiting value of this cross-partial, as V_j approaches zero, is

$$\omega_{ji}[1 - P_j \ln P_j] = \omega_{ji}[1 + \text{ShannonEntropy}_j]. \qquad [48]$$

This is quite intriguing. If weight is unity, then, as affect is dialed down, the impact of j's affect on i's disposition approaches 1 plus the binary entropy of j's probability distribution. If j is certain, this entropy is zero and the partial is just ω_{ji} as in [45].[167] But otherwise, j's estimate affects i's disposition in an information-theoretic way. Again, this result is independent of the specific functional forms chosen for V and P.

Political Corollary

Observe also that entropy, $-p \ln p$, is hump-shaped (unimodal) with minima of zero at $p = 0$ and $p = 1$. So, to maximize her marginal effect on i's disposition, j should, in fact be "evenhanded," expressing a probability

[166] Here, we return to the form without any imputation of thresholds. As the latter are constants, they do not affect the derivatives of interest here. Neither do the thresholds; this derivative will be identical to that computed on net disposition.

[167] If $P_j = 0$, the entropy is 0 by convention, since technically the $\ln P$ term goes to negative infinity as P approaches 0.

FIGURE 60. Stationary Agent in Stimulus Field

estimate of 0.5. If politicians knew this theorem, political debate might be far less polarized.

Equation [48] is, as noted, a limiting case. But, precisely such an asymptotic connection to entropy would be hard to notice from agent simulations alone, and serves once more to illustrate the usefulness of having mathematical and computational versions "talk to one another."

Now let us return to our agent, situated in a spatial field with aversive stimuli, as shown in Figure 60.

We immobilize our subject and bombard her with aversive orange stimuli. With no affective influence, her increasing fear has no effect on her estimate of orange relative frequency (probability), as shown in Figure 61.

However, if affect/emotion influences probability judgment as specified in equation [44], then probability escalates under the *identical* affective trajectory,[168] as is shown in Figure 62.

[168] Ceteris paribus, the same random seed (here equal to 1) will produce identical runs.

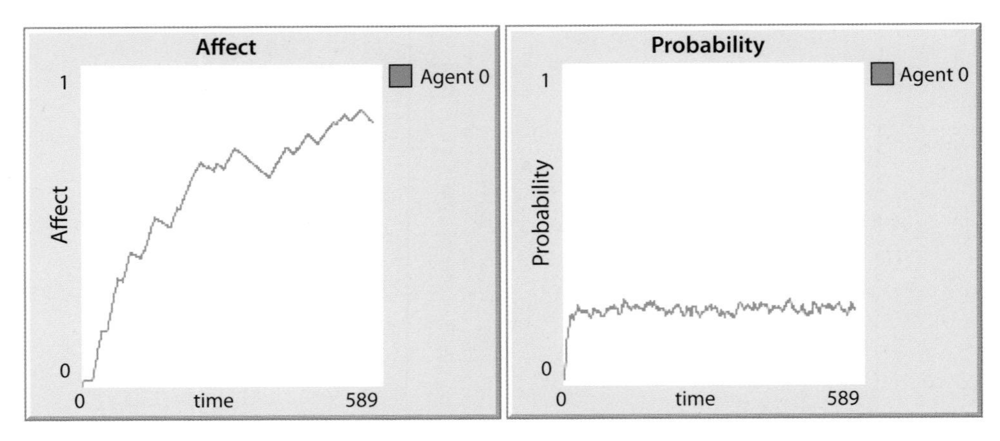

Figure 61. No Affective Amplification

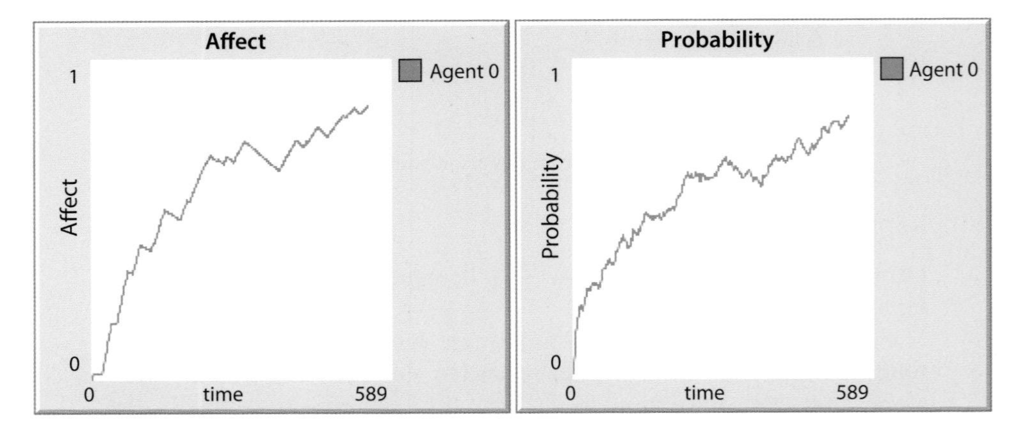

Figure 62. Affective Amplification

Mutual Amplification

The individual's dispositional dynamics (since it varies as the sum of affect and probability) will be amplified accordingly, as will the agent's level of action, be it retaliatory violence, pandemic flight, or financial panic. Under this kind of emotional amplification, the network of *coupled* agents can easily work itself into a mutually reinforcing dispositional frenzy, with precious little hard empirical evidence, as illustrated by the low initial *P* curves of Figure 59. Again, *this is far from canonically rational behavior, but is well within the behavioral compass of Agent_Zero society.*

Zillmann's Experiment Revisited

We discussed the general interaction of emotional arousal and cognitive/evidentiary inhibitors of retaliation earlier in connection with the experiment of Zillmann et al. (1975). Zillmann does not propose a mathematical relation. But the one we've introduced is broadly consistent with the pattern he observed. Specifically, in our development thus far, we have been interpreting P as the empirical component of our disposition *to* act. Hence, we can interpret $(1 - P)$ as the mitigating evidence against action. Previously, we postulated that the P-value is amplified by emotional arousal (e.g., fear) as

$$P_e = P_n^{1-V}. \quad [49]$$

But then the *mitigating* evidence, $1 - P$, is damped accordingly: as the level of arousal (V) increases, the weight placed on mitigating evidence $(1 - P^{1-V})$ decreases. In the limit of maximum V, it is ignored entirely, since in that $V = 1$ case, we have

$$1 - P_e = 1 - P_n^{1-1} = 1 - 1 = 0. \quad [50]$$

as per Zillmann et al. (1975). Here, then, is a candidate mechanism that could, in principle, be tested in a laboratory.

Group Implication

Turning as always to the group implication, we observe that *if everyone is subject to affective amplification as before and their dispositions are strongly coupled (high weights), then extremely explosive dispositional dynamics can arise from stimuli that would not suffice to trigger any individual's isolated action.*

Agent_Zero Does Jury Duty

This problem is recognized in the rules of evidence in jury trials. It is understood that emotionality—high V produced by inflammatory evidence—can bias a juror's estimate of the guilty probability (P_e). Moreover, the present model would suggest that with sufficient weight, the biased opinion of even one juror could, through contagion, produce a majority guilty verdict, where no juror operating alone would render one. For evidence of powerful conformity effects in jury processes, see Sunstein and Hastie (2008); Hastie (1993); and Hastie, Penrod, and Pennington (1983). We will model a jury trial shortly. To do so, we need a model of endogenous weight dynamics.

III.7. ENDOGENOUS DYNAMICS OF CONNECTION STRENGTH

To this point, interagent connection strengths—the weights—have been exogenous constants assigned at the start of each run. We now wish to generalize this picture in several respects. To begin, we introduce dynamic connection weights by making connection strength an endogenous function of affective similarity (or *homophily* if you prefer). We suggest how social media can function to amplify these affective dynamics and ultimately embolden political resistance, as seen in the 2011 Arab Spring, a "toy" version of which is generated. Obviously, parables of emotionally charged health behaviors (e.g., fear of autism and vaccine refusal) and other collective phenomena can be generated as well. One could, of course, make connection strength depend on cognitive rather than emotional affinities. And we briefly discuss how this also can be done in the framework. But we start with affective homophily.

Affective Homophily

"Birds of a feather flock together." This is *homophily*, the tendency to associate most closely with those most similar to oneself. Homophily is an important phenomenon in explaining network formation and the dynamics of network structure. It is very important from a policy standpoint. Suppose you observe that some bad habit is prevalent in a certain social network and that you wish to reduce its prevalence. The effectiveness of various policies will hinge on whether (a) people "caught" the bad habit through contagion on the network, or (b) they joined the network in the first place because they wanted to be with habituated others. So, is observed prevalence the result of contagion or homophily?[169] If the mechanism is contagion, then "busting up" the network may affect prevalence by blocking spread. But if homophily was the mechanism of network formation, there is no spread to block, and the same strategy will have no effect. Indeed it could even backfire, as when military attacks on a social network simply increase antiattacker sentiment, boosting recruitment through affective homophily. We begin by exploring a simple picture of affective homophily, taking up other types of homophily below.

Considering Agent i and Agent j, suppose we define their affective homophily (h_{ij}) as 1 minus the absolute value of the difference between their affects. Here, we don't care whose affect is larger or smaller; we care only about the size of the gap, hence the absolute value. But, we want the measure

[169] This is a statistically complex matter when dealing with data (Shalizi and Thomas, 2011). Of course, we know the mechanism is homophily here.

to be maximal (i.e., 1.0) when the difference is zero. So, for the $h_{ij}(t)$ we use 1 minus the unsigned affective difference[170]—that is,

$$h_{ij}(t) = 1 - |v_i(t) - v_j(t)|. \qquad [51]$$

Using this alone as the weight of i on j (and vice versa) would have two shortcomings. First, it ignores magnitudes. If two agents feel absolutely nothing (both have affect of zero), their connection weight will be 1, exactly as if they had equal but maximal affect. One would assume that the latter passionate pair would form a stronger bond than two literally indifferent nudniks. A second problem with simply using h_{ij} as the weight is that, assuming runs begin with everyone at zero affect, this functional form forces all initial connection weights to be 1.0, which seems amiss. A remedy to both problems is to premultiply equation [51] by the sum of the affective magnitudes, $v_i + v_j$.[171] Then weight is given by

$$\omega_{ij}(t) = (v_i + v_j)(1 - |v_i(t) - v_j(t)|). \qquad [52]$$

This has the properties we want: (1) it distinguishes the case of equal and passionless from that of equal and passionate, through the magnitude term, and (2) in runs beginning with zero affect, the weight begins at zero rather than 1.[172]

Weight Surface

As a function of v_i and v_j, we see the ω_{ij} surface in Figure 63. It has two distinctive features. We see a ridge rising from 0 to 2 as v_is are held equal and increased from 0 (passionless) to 1 (passionate). But we also see that emotional polarization reduces weight, which is also 0 when $v_1 = 0$ and $v_2 = 1$, for example.

Using equation [52], the final model will actually give crude, but I think novel, answers to the basic questions: How do networks change? Why do networks happen? Or, speaking more technically:

1. How may weights change endogenously?
2. How can these underlying continuum dynamics generate network *structure* proper (the binary formation and dissolution of edges)?

[170] Notice that there are only three links because the functional form imposes the symmetry $h_{ij} = h_{ji}$. The same will hold for weights, ω_{ij}.

[171] One could normalize by 2 to bound weights to the unit interval, but at this scale it's cleaner not to.

[172] As earlier noted, it is clear that $\omega_{ij} = \omega_{ji}$.

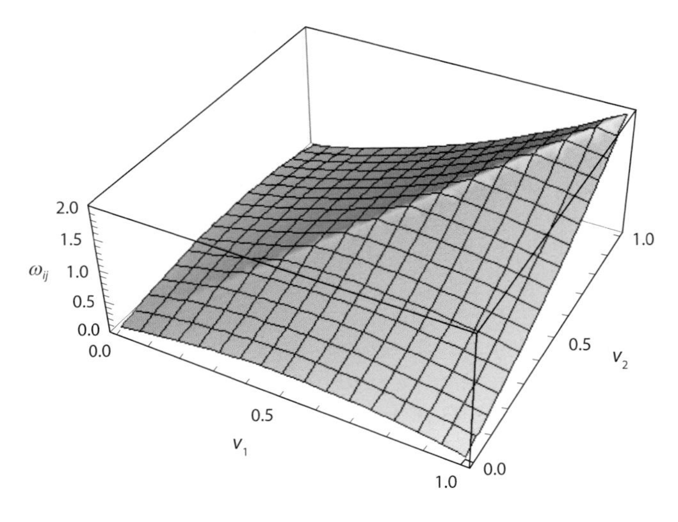

FIGURE 63. Weight Surface

Just to suggest the range of dynamics—first excluding space, and then reintroducing it in the agent version—we toss the ball back to differential equations and offer three runs. The first uses the classical Rescorla-Wagner model for $v(t)$. The second introduces different exponents (the δs, introduced in equation [24] of Figure 23, and the third adds heterogeneous learning rates (α, β). Throughout, the initial value of v (affect) is 0 for all agents, and for all agents, λ (the maximum associative strength) is 1.0. The probabilities (P) are identical and are an arbitrary constant. Having reviewed the behavior of this purely mathematical version, we then introduce space and the agent model and show how fundamentally different behaviors emerge. All differential equations and their numerical *Mathematica* solutions are given in Appendix II and the *Mathematica* Notebook.

General Setup

The central departure, of course, is in making the weights dependent on affect. Substituting expression [52] for each weight into our skeletal disposition formula for Agent 1, we obtain

$$\begin{aligned} D_1^{\text{net}}(t) &= v_1(t) + p_1 - \tau_1 \\ &+ \{(v_1(t) + v_2(t))(1 - |v_1(t) - v_2(t)|\}(v_2(t) + p_2) \\ &+ \{(v_1(t) + v_3(t))(1 - |v_1(t) - v_3(t)|\}(v_3(t) + p_3). \end{aligned} \qquad [53]$$

The bracketed term is simply our new affect-dependent weight. Analogous expressions can be obtained for the other agents. Notice that, while this looks more complicated than the original model with weight an exogenous constant, it actually endogenizes weight, eliminating $N(N-1)$ freely adjustable parameters for each population size N. Now for some illustrative runs.

Case 1. Classical Setup

In the classic Rescorla-Wagner setup, δs (introduced in equation [24], Figure 23) all equal zero. As a first run, we will impose this homogeneity. As in the earlier mathematical development, we imagine a continuous stream of trials. Agents will also have the same α and β, and P-values are identical. This of course dictates that all dispositional, affective, and weight trajectories are identical across agents. They are displayed in Figure 64 for an action threshold of 1.5.

The upper (common) curve plots D^{net}, disposition minus threshold. It begins negative, then crosses the action threshold (zero in the net-of-threshold form) at roughly $t = 2$ and rises with increasing affect. Affect is the lowest curve and rises from zero to its maximum of 1.0 for each agent. The middle curve is weight. At the outset, there is no network at all—weights are zero. But with homophily maximized (since affects are equal), weight increases with the sum of paired affective magnitudes, topping out

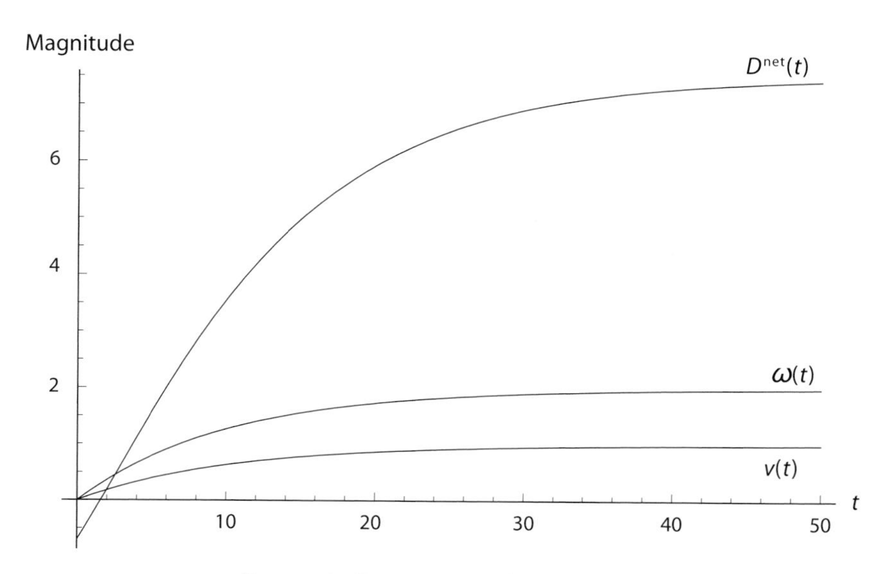

FIGURE 64. Homogeneous Classical Agents

at 2.0, as expected. *We go from no network to a maximally weighted one by the mechanism offered here.* Now, let us introduce heterogeneities

Case 2. Heterogeneous Exponents

Leaving everything else exactly as it was, recall the generalized Rescorla-Wagner model, equation [24] of Figure 23. The generalization introduced a parameter δ controlling the "S-curviness" of the associative strength function $v(t)$. For $\delta = 0$, we get the original Rescorla-Wagner model (which is always concave down). Here, let us assume that each agent has a different δ. Assume that $\delta_0 = 1$ and $\delta_1 = 0.8$, making them (distinct) S-curve learners, while $\delta_2 = 0$ (classical). Dispositional dynamics are quite different across agents, as shown in Figure 65.

Although the disposition curves ultimately increase, each changes concavity multiple times. The weight trajectories for this same run, shown in Figure 66, are of central interest, showing an ebb and flow of connection strength finally converging to a single maximum. Notice that all weights begin at 0 (no network) and that the green relationship between Agents 0 and 1 nearly collapses at $t = 50$ but recovers and ends strong.

The weight trajectories are clearly nonmonotonic. However, they are provably convergent to a unique equilibrium value of 2λ. Specifically, while the dispositional dynamics are coupled, the affective dynamics proper [the $v(t)$ equations] are not. Each converges to its maximum associative

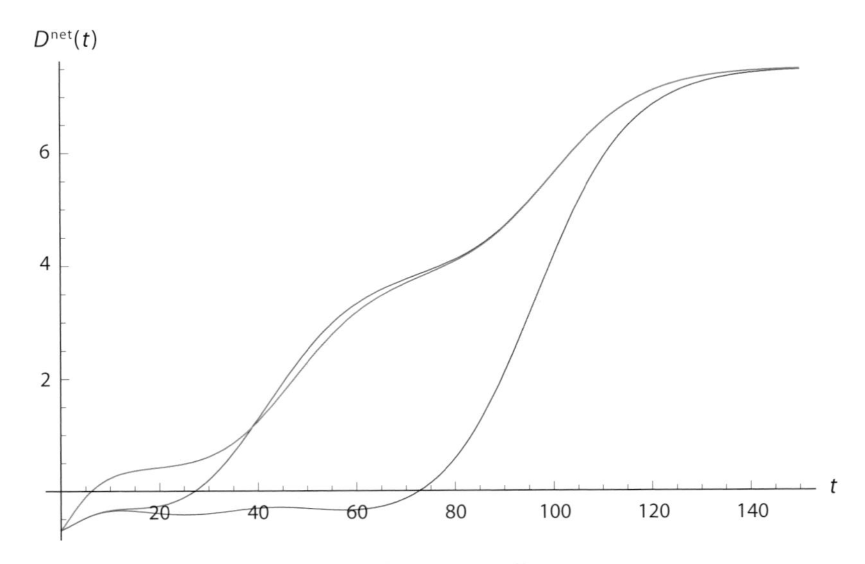

Figure 65. Heterogeneous Exponents

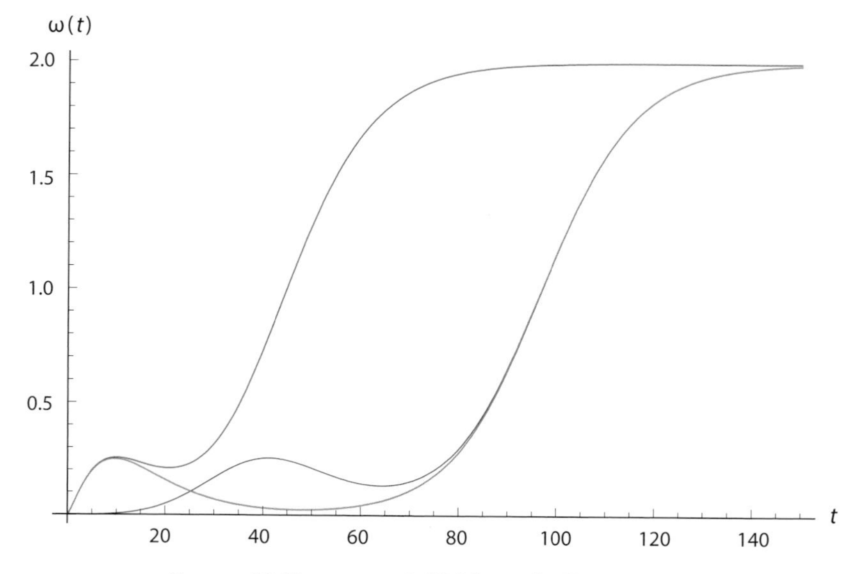

FIGURE 66. Nonmonotonic Weights under Heterogeneity

strength, here the common value, λ. In the limit,[173] therefore, all affects are equal, and so the affective differences are all zero. Hence, the bilateral weight expressions (equation [52]) all reduce to the sum of paired affects. But, as just noted, these are both equal to λ, so their sum is 2λ, as claimed.[174]

In general, each agent i could have a different λ_i. Then, for every ij, the equilibrium weight is

$$\begin{aligned} \omega_{ij}(\infty) &= (v_i(\infty) + v_j(\infty))\,(1 - |v_i(\infty) - v_j(\infty)|) \\ &= (\lambda_i + \lambda_j)(1 - |\lambda_i - \lambda_j|), \end{aligned} \qquad [54]$$

which obviously reduces to 2λ in the preceding case.[175]

Affinity Trajectories

At any time, there are three interagent weights defining a point in 3-space. So, over time, they trace out a particular space curve. One might term this the group's *affinity trajectory*. For the weights just plotted, this affinity

[173]The limiting value as t approaches infinity.

[174]Normalization to eliminate the 2 introduces unnecessary clutter at this 3-agent scale.

[175]$x(\infty)$ denotes the limit, as t approaches infinity, of $x(t)$.

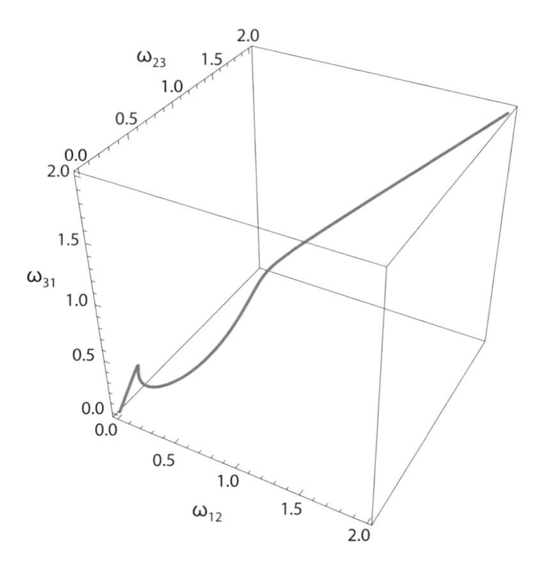

Figure 67. Affinity Trajectory

trajectory is shown in three dimensions in Figure 67. It converges to the point $(\omega_{12}(\infty), \omega_{23}(\infty), \omega_{31}(\infty))$. If we assume a positive Rescorla-Wagner affective extinction rate (discussed in Part I), this equilibrium point is the origin.

Case 3. Heterogeneous Exponents and Learning Rates

The out-of-equilibrium dynamics get more complex if we now add to the preceding different learning rates, the α's, and β's. If we simply call their product k, and set k_1, k_2, and k_3, respectively, to 0.01, 0.04, and 0.08, the group's affinity trajectory is very different, including a sharp hysteresis, as shown in Figure 68.

We will discuss link thresholds fully later. But, to anticipate slightly, let us assume—as seems defensible—that when the affective difference between two agents is sufficiently close to 1.0 (say 0.95) the link between them is broken. Under this interpretation, all these diagrams tell stories of endogenous changes in network *structure* (the pattern of links per se). Links dissolve, and then are re-established as affective differences rise and fall.

If we now move to the agent world and add a spatial distribution of stochastic aversive stimuli, the affinity and structural dynamics are more complex still and *no equilibrium is assured*. This is a central difference between

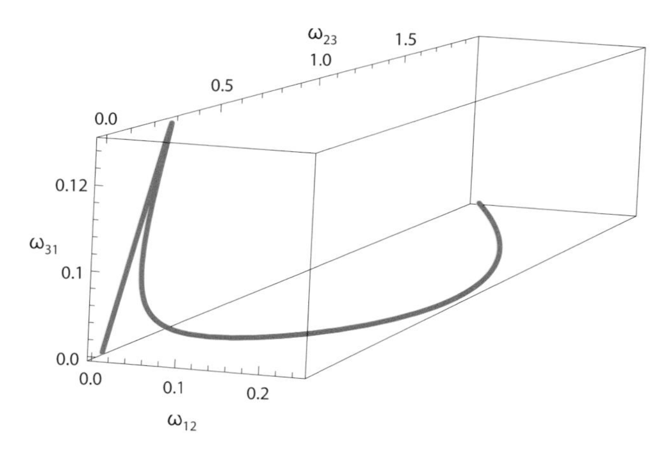

Figure 68. Affinity Hysteresis

the mathematical and agent models. A single example will suffice to illustrate the point.[176]

Agent-Based Model: Nonequilibrium Dynamics

In fact, as we will see, in the spatial agent version, weight trajectories need *not* converge. Here is another example in which the mathematical and agent versions enrich one another. Without the mathematical version, one might not notice the convergent tendency in the first place. But with the mathematical version alone, you might never discover that divergence is possible. The agent model exhibits this. With both mathematics and agent-based modeling, you just learn more.

Case 4. Spatial Agents

To see the affinity dynamic in action, I present a familiar case. The run starts with all three agents in their initial positions before any attacks, with all connection weights initially equal to zero. Then attacks begin. These increase the v-values of the agents and, in turn, the interagent weights, which grow in the crucible of combat. *The thickness of each link will equal the weight of that link.* So, as these weights change through our affective

[176]Other weight-adjustment schemes are suitable for a longer study.

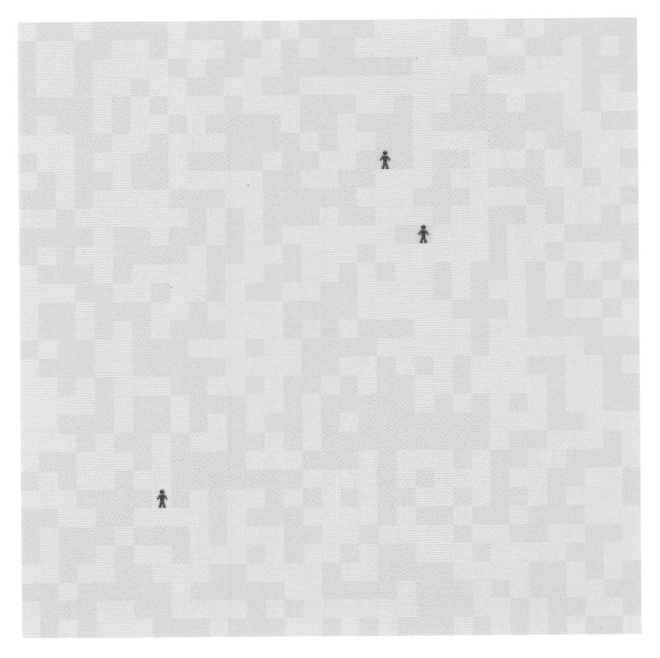

FIGURE 69. Initial Weights Zero [Movie 9, start]

homophily dynamics, so will the thickness of these links. Initially, the picture is as shown in Figure 69.[177]

As things unfold, the experiences and affective trajectories of the agents begin to diverge. And the network structure encoded in the weights evolves accordingly, as shown ($t = 500$) in Figure 70.

Agents 2 and 3 develop the closest relationship, followed by that between Agents 2 and 1. Finally, when the entire yellow population has been annihilated, the agents' pairwise relationships are not what they were before the war, as shown in Figure 71.

At this point, all aversive (Orange) stimulation has stopped and with it, any further change in weight. The entire evolution is recorded in **Movie 9,** on the Princeton University Press Website.

This equilibrium is sensitive to extinction rates. Here, affective extinction was zero. If the extinction rate is positive, then each agent's affect eventually damps to zero; at this point their affective strengths are zero, as in turn are their weights, and the agents go their separate ways in life.

[177] Numerical values used are given in the Appendix IV table of assumptions and are also available from the Source Code in the Applet.

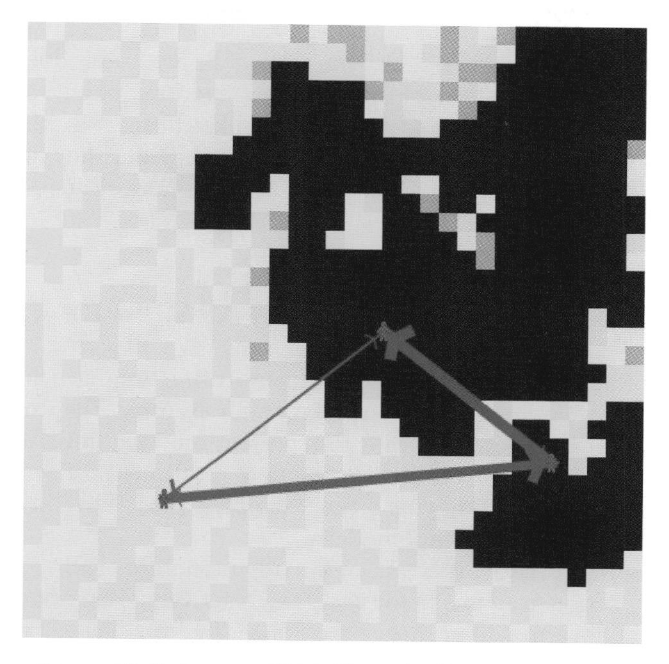

FIGURE 70. Endogenous Weight Dynamics [Movie 9, continued]

FIGURE 71. Endogenous Weight Dynamics [Movie 9, terminus]

As noted earlier, a general model should generate not only dark parables of baseless massacre but also parables of hope. And it does. An example unfolding at the time of this writing is the wave of anti-autocratic revolutions in the Middle East. A salient aspect of this so-called "Arab Spring" is the unprecedented central role of social media and endogenous networks in the dynamics.[178] The *Agent-Zero* framework generates this parable also.

III.8. GROWING THE 2011 ARAB SPRING

Not all antiautocratic revolutions eventuate in functioning democracies. The Iranian revolution installed a theocracy, for example. The variant I am about to show, therefore, is meant to represent only the initial resistance and overthrow phases of a stylized revolution: the Tunisian, Egyptian, and Lybian revolutions of 2011 being recent examples. The long-term evolution of these political systems is not modeled.

Here we interpret the space as an initially monolithic ruling authority. The binary action available to the agents is to actively resist or not. Yellow squares are government entities—the police, the intelligence apparatus, government-controlled media, and so forth. Now we interpret orange activations as instances of government corruption (e.g., nepotism, bribery, abduction, human rights abuses). These are the stimuli that increase the agent's affect $V(t)$ against the government—by our usual Rescorla-Wagner mechanism. In addition, agents are sampling within their vision to estimate the relative frequency of orange—the probability $P(t)$ of a randomly selected government entity being corrupt. Let us refer to the sum of these as the agent's grievance. But, grievance alone does not a Molotov cocktail make. What is the *risk* of open resistance? The threshold term represents this.

If even the slightest expression of dissent will result in execution, the deterrent threshold τ is very high. On the other hand, if the direct open exposure of government corruption is tolerated, then the threshold is low. So, we can think of one's total disposition as grievance minus threshold (similar to the Epstein Civil Violence Model[179]). But this is just our standard skeletal formula:

[178] This is not to deny an important role for the "social media" of the day in earlier bottom-up revolutions. Samizdat in the USSR, Cato's Letters in the American Revolution, and audio cassettes of Khomeini speeches in Iran would be examples.

[179] J. M. Epstein (2002).

$$D^{\text{net}} = V + P - \tau. \quad [55]$$

Of course, the entire point of suppressing freedoms of assembly and speech is to keep people in isolation. In the Arab Spring, social media played a huge role in thwarting these classic repressive measures, allowing networks to form and weights to increase to the point where dispositions to protest exceeded thresholds and the state was swept away in a flood of resistance (formerly dark red, now jasmine colored). Social media certainly did not *create* the underlying grievance. But, through dynamic network effects, it powerfully facilitated its amplification, emboldening the ultimately victorious rebels.

Let us generate this result in the *Agent_Zero* framework, beginning with no social media.

No Social Media

The left frame in Figure 72 shows our agents at $t = 0$. The right frame shows them in a sea of government corruption. With no social connection or reinforcement, the isolated Blue agents simply endure the ongoing corruption; there are no jasmine outbursts of resistance.

Although their antigovernment affect rises to its maximum possible levels (right frame of Figure 73), solo dispositions (since they are isolated)

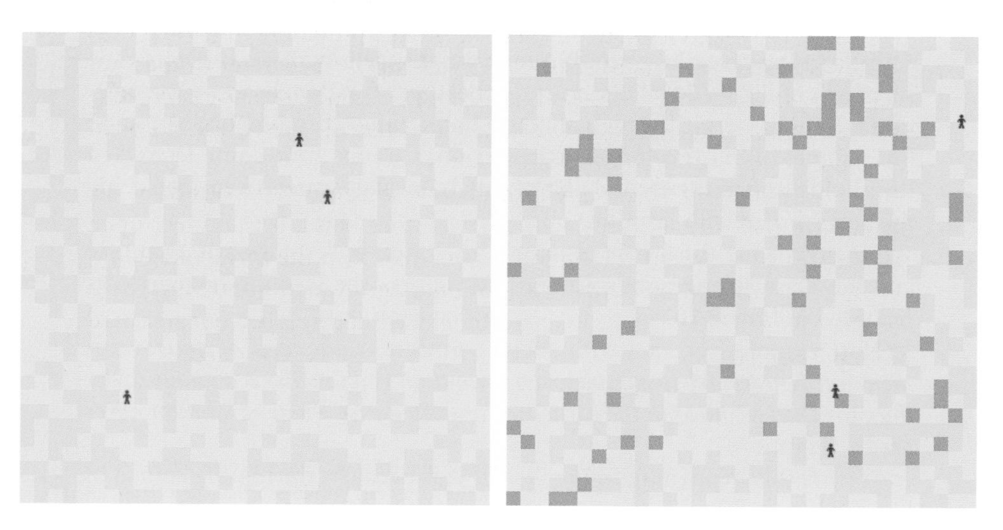

FIGURE 72. No Social Media

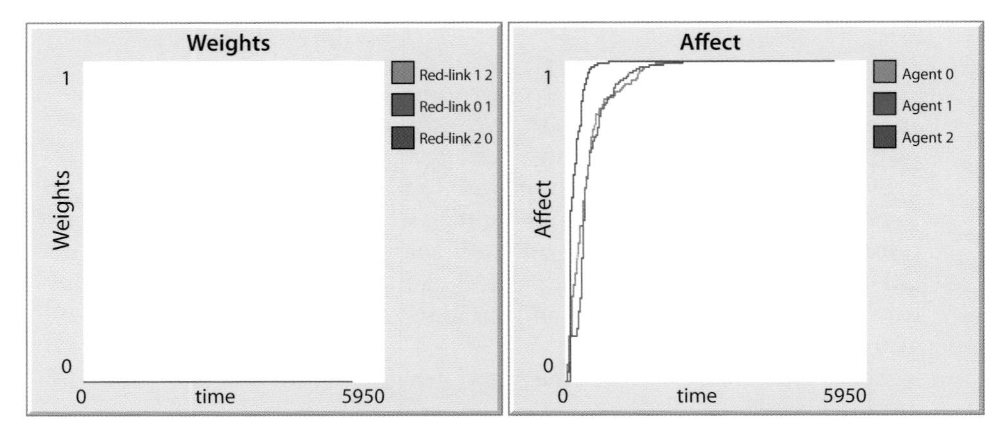

FIGURE 73. Weights and Affect with No Social Media

to openly resist do not exceed the risk threshold, and no rebellious action occurs. In sum, with all connection weights zero (left frame of Figure 73), no individual's disposition can amplify any other's. And despite antigovernment affect rising to maximal levels, no rebellious action occurs.

Endogenous Weight Adjustment through Social Media

If we now introduce the earlier apparatus of endogenous network weight adjustment, a radically different story unfolds. All other parameters are exactly as in the preceding run. But now weights can increase as before. The six panels of Figure 74 tell the story. In the upper two panels, we see connection weights growing through affective homophily (common opposition to the regime). *Total* (no longer isolated) dispositions thus come to exceed thresholds, and in the second pair of panels, local uprisings occur, replacing the government with jasmine.

Finally, in the third pair of panels, the jasmine revolution is complete and there are strong bonds among the victors. The jasmine revolution is recorded in **Movie 10**.

Notice, moreover, that with social connection, *the level of antigovernment affect necessary to complete the revolution* (Figure 75) *is actually lower than the level attained in the earlier unconnected case where revolution never occurs* (Figure 73).

Why is this happening *in the model?* It is because, during the course of the revolution, Orange stimuli (instances of corruption) are being overwritten (rooted out and eliminated) by jasmine patches (the damage radii), so

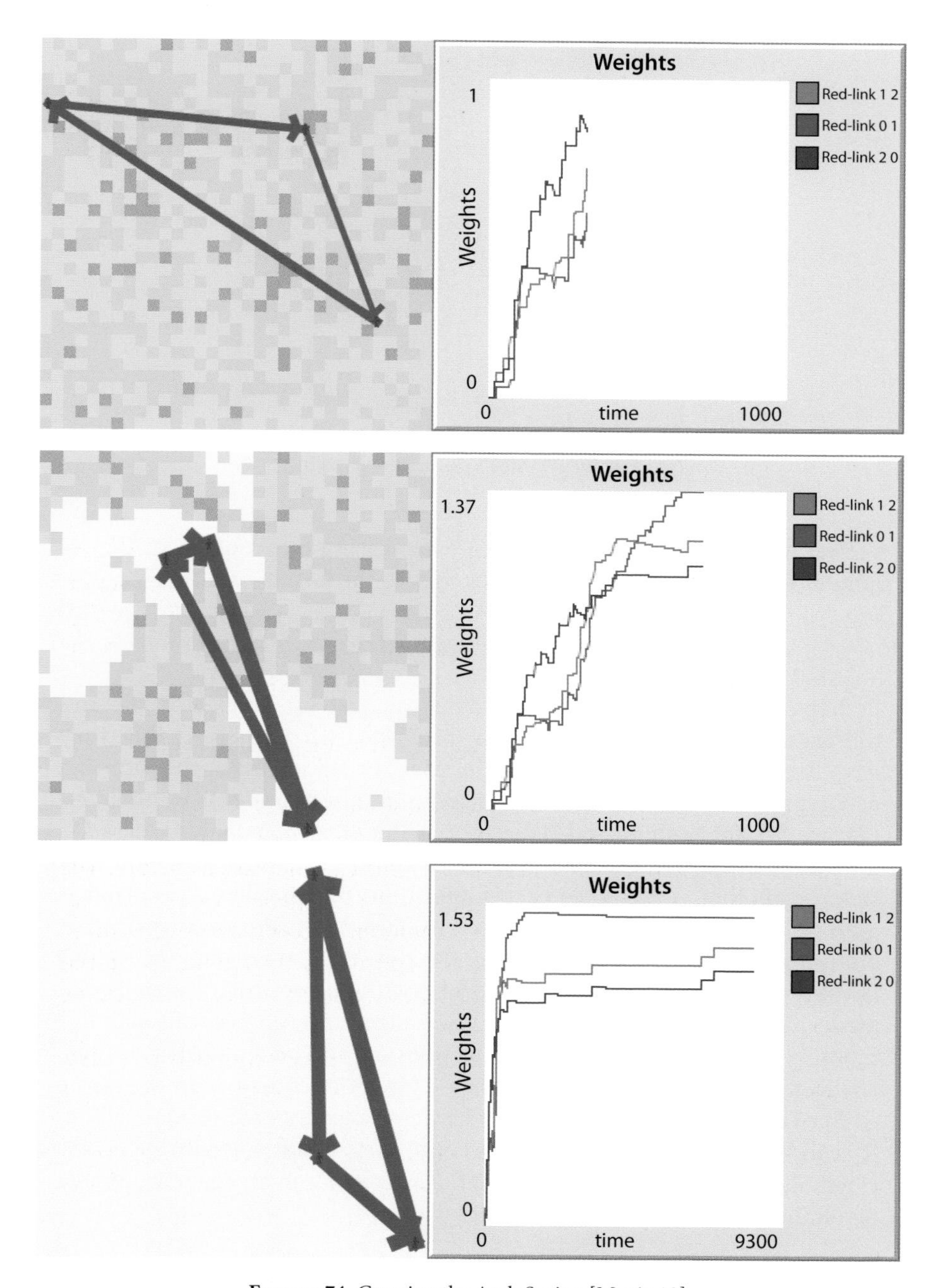

FIGURE 74. Growing the Arab Spring [Movie 10]

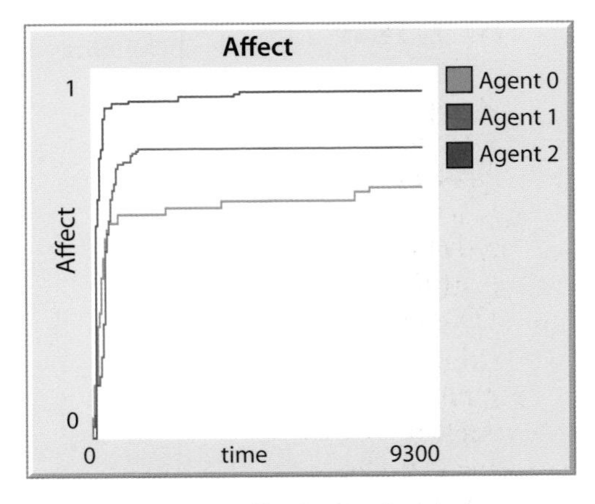

FIGURE 75. Affect During Revolution

the stimuli (the conditioning trials) are decreasing in frequency as the regime is eradicated. Hence affect never rises to the former level![180]

Revolt of the Swarm

Perhaps the most remarkable thing about the Arab uprisings of 2011—and the feature that sets network effects into such sharp relief—is this: they have been *leaderless*. Name the Lenin, Mao, or Khomeini of the Tunisian or Egyptian Revolutions. Though ongoing at the time of this writing, the Libyan, Syrian, Yemeni, and Bahraini rebellions are equally leaderless. As if to echo Tolstoy, Engels famously remarked that "If Napoleon had been lacking, another would have been found." But here, no alter-Napoleon is even necessary. It is truly the network—the swarm—that is making the revolutions. At the time of this writing, Occupy Wall Street demonstrations have spread widely around the world, also with no particular leader. Social media have made this possible, permitting affective (even dispositional) homophily to strengthen ties from the bottom up, and on a global scale. This is a fact fully appreciated by repressive regimes, of course, who—in many Arab countries—shut down Internet service to prevent organized rebellion. It is the Internet equivalent of a physical curfew. But freedom of cyber-assembly is a portentous development. It may usher in an age of leaderless revolutions.

[180] This mechanism is reminiscent of the fight-vs.-flight example, in that stationary fight eliminates stimulus, whereas flight requires passage through a stimulus-rich environment. When actions endogenously change the environment, counterintuitive results may occur.

III.9. JURY PROCESSES

In revolutions, existing legal systems are often destroyed. But the same *Agent-Zero* framework can be used to model legal systems in operation. In the examples thus far presented, emotional, deliberative, and social forces operate concurrently. In a trial by jury, certain of these are sequenced. Most distinctively, social interactions *within* the jury proper are forbidden until the jury is sequestered. At this point, powerful social effects are known to operate, driving jurors to conform to a majority (recalling the Asch experiment) as it unfolds behind closed doors. Of course, a complex emotional and deliberative evolution has already taken place for those who are empaneled. We wish to model three phases in the *Agent_Zero* framework: the pretrial public phase, the courtroom trial phase, and the jury deliberation phase. Each will be assumed to last 30 days.

Phase 1. Public Phase

Imagine a celebrity trial such as the O. J. Simpson case. Long before there was a courtroom trial, there was a vibrant public discussion of O. J.'s guilt. Let orange patches represent assertions of O. J.'s guilt. Yellow patches take the default position—innocent until proven guilty. Imagine three Blue agents who do not know one another and who have no idea that they are fated to become jurors. As they go about their business in the landscape of public opinion, they are bombarded with stimuli regarding the impending trial. In Figure 76 they are shown positioned randomly in the public space. The courtroom is the gray square that is empty in this phase of the process. One can imagine different neighborhoods with different modal attitudes toward O. J.

As usual, these stimuli are processed by our Blue agents. Their affect V toward O. J. is being driven by the flow of orange stimuli, just as before. And they are recording the ratio P of orange to observed patches as the crude empirical measure of O. J.'s culpability.

Employing the memory apparatus developed earlier, we assume that each Blue agent has a memory of 90, meaning that her most recent 90 probability estimates, or impressions, if you prefer, are stored.[181] We also assume that the moving average over this sample is employed in forming her overall disposition regarding O. J.'s guilt. The P used in computing each agent's disposition, in other words, is this moving average.

[181] In this 90-day process, therefore, all probability estimates are remembered.

FIGURE 76. Landscape of Public Opinion [Movie 11]

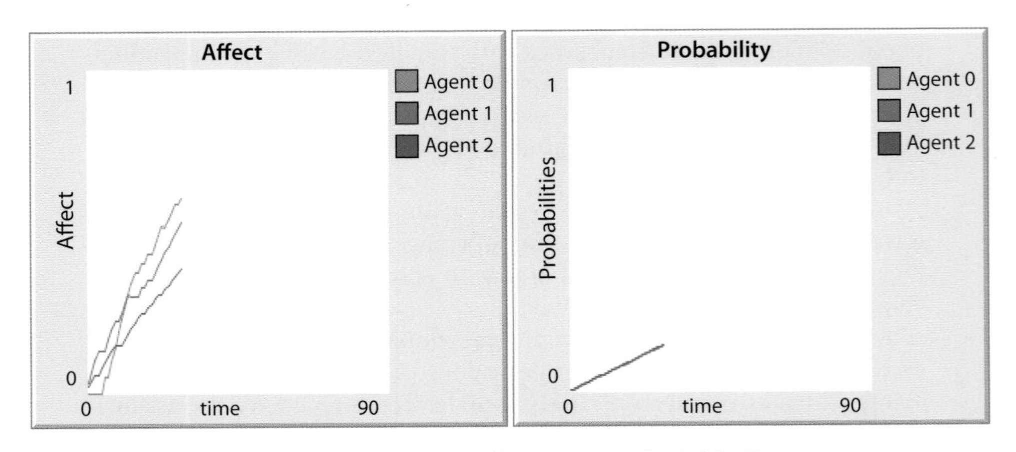

FIGURE 77. Affective and Deliberative Dynamics, Public Phase

These Phase 1 affective and deliberative components are plotted in Figure 77. Agents are exposed to effectively identical activation patterns, so their probability estimates are identical. But, because they have heterogeneous learning rates, their affects differ. These (minus their identical thresholds) are combined to produce in each agent a net disposition to pronounce O. J.

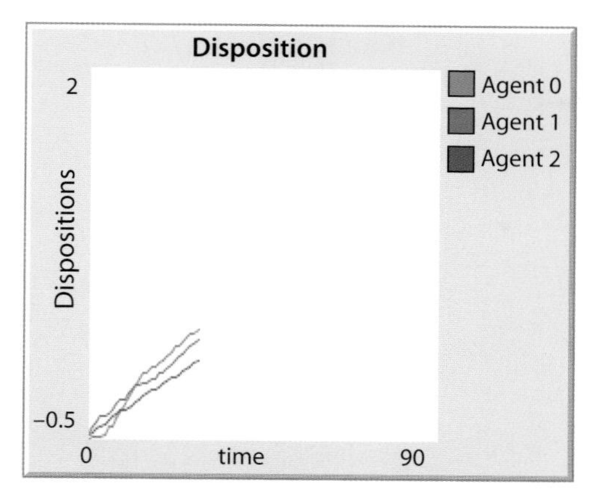

FIGURE 78. Disposition to Convict Dynamics, Public Phase

guilty if asked. If this exceeds zero, they vote guilty; otherwise, innocent. Of course, this changes over time, as shown in Figure 78. At this point none of our jurors-to-be believe Mr. Simpson to be guilty.

Importantly, interagent weights are zero in this phase. The agents are still complete strangers. This is about to change.

Phase 2. Courtroom Trial Phase

We imagine that our three agents are chosen to become the jury in the courtroom trial. Now the venue changes from the court of public opinion to the courtroom proper—the upper-right square. Previously, this was an empty gray, while the public area was alight with orange and yellow assertions. Now these landscape colors are reversed. As in Phase 1, interagent weights remain zero, since technically, jurors do not communicate in this trial phase. However, it is no longer the barber, the coworker, or the pundit who is offering orange and yellow stimuli. It is the lawyers for the prosecution and defense. Pursuant to rules regarding evidence, for example, the pattern of stimuli can change abruptly from Phase 1. Indeed, where Phase 1 may have generated a social majority opinion of guilt, evidence presented in court may favor the defense. Yellow (innocence presumptions) indeed outnumbers orange in the courtroom proceeding, as depicted in Figure 79.[182]

Whether this sways the jurors from their pretrial dispositions depends on a number of factors represented in the model.

[182] Hence, the relative frequency of orange is lower than in Phase 1.

FIGURE 79. Jury in the Courtroom [Movie 11, continued]

Extinction

Try as they may to suppress it, jurors often carry their pre-trial affect into the courtroom. For some, emotions fade. In the language of the Rescorla-Wagner model, extinction occurs. In others it does not. Jury selection tries to weed out candidates who have an obvious emotional investment in one verdict over another. But, as we have already discussed at length, people may not be aware of their actual emotions, and despite appearing impartial under questioning, they may well harbor deep unconscious associations and presumptions of guilt or innocence. We can explore the effect of low and high extinction rates on the affective levels in place when the trial phase starts. Continuances may provide a time window in which some extinction can occur.

Memory

In the public pretrial phase, the agents will have formed probability judgments as well. Using the memory apparatus developed earlier, we imagine the jurors carrying forward a moving average of their estimates (relative frequency of orange within vision) over a memory window. For those with high memory, these pre-trial probabilities will be insensitive to new data for a number of periods. Those with short memory will exhibit more rapid adaptation to the new courtroom evidence. For expository purposes, we will assume that all agents have memory 90 and that the moving average (rather than the moving median) over this window is the *P*-value used in computing each agent's disposition to convict.

How much change occurs in these affective and probabilistic components also depends on the duration of the trial phase. As noted, I have divided the pretrial, trial, and jury phases equally at 30 periods. But, as in the real world, these can be quite arbitrary, and continuances and changes of venue may affect the course of events. The latter is explored later.

Phase 3. Jury Phase

Finally, when all testimony and evidence for the prosecution and defense have been presented, the jury is instructed and sequestered in the jury chamber to arrive at a verdict. Now, for the first time, the jurors interact directly with one another. The social dynamics within juries can be very complex, with subnetworks vying for unattached jurors, who may join by affective homophily, or agreement on other dimensions. Here, we will focus on affective homophily, as modeled earlier. Just as before, inter-juror weights are given by the product of strength and homophily:

$$\omega_{ij}(t) = (v_i + v_j)(1 - |v_i(t) - v_j(t)|). \qquad [56]$$

And, as before, each juror's net disposition to convict is given by

$$\begin{aligned} D_1^{\text{net}}(t) &= v_1(t) + p_1 - \tau_1 \\ &+ \{(v_1(t) + v_2(t))(1 - |v_1(t) - v_2(t)|\}(v_2(t) + p_2) \\ &+ \{(v_1(t) + v_3(t))(1 - |v_1(t) - v_3(t)|\}(v_3(t) + p_3). \end{aligned} \qquad [57]$$

If we depict the jury chamber as a small brown (oak-paneled) room, the picture of interagent weights and the sharp potential effect on verdicts is illustrated in Figure 80.

Some pairs are more closely bound (thicker links) than others, and the effect of weights can be huge, as shown by the sharp dispositional jump they induce, compared to the dispositions the jurors had before this phase. Weights are "off" until this phase but jump abruptly when affective homophily is discovered and bonds are formed, as depicted in the weight plots of Figure 81.

Figure 80. Jury Chamber [Movie 11, terminus]

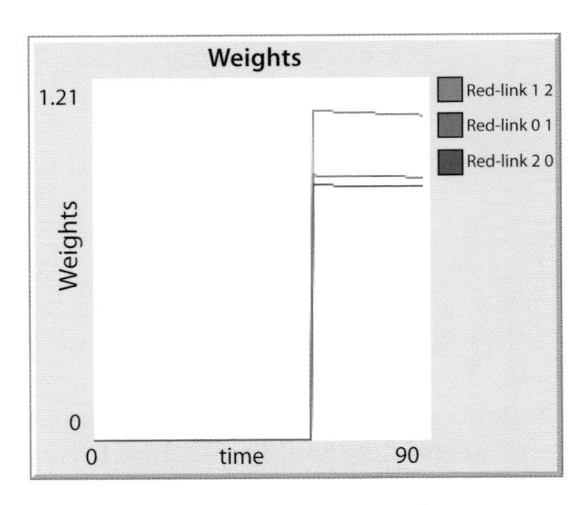

Figure 81. Interjuror Weights

This can eventuate in sharp changes in the disposition to convict, compared to earlier phases, as illustrated in Figure 82.

Indeed, *no jurors would have convicted before this jury phase, but they are unanimous in rendering a guilty verdict, having interacted directly*. The entire history is recorded in **Movie 11**.

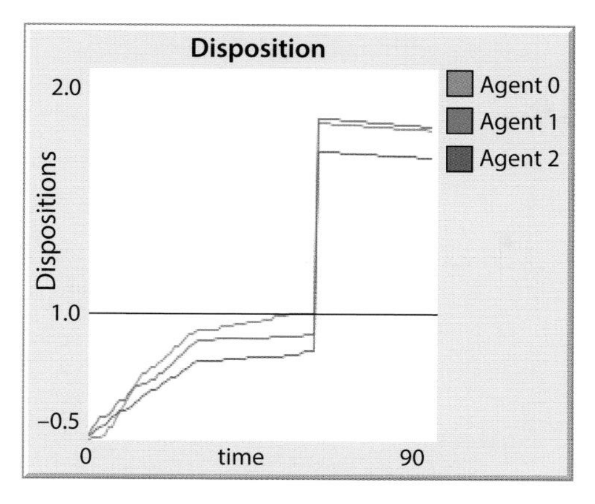

FIGURE 82. Jump in Conviction Dispositions through Endogenous Weights

The rationale for various well-known legal tactics emerges naturally from the framework. One of these is a change of venue.

Change of Venue

If a judge deems a fair and impartial jury to be impossible in a particular location, a change of venue may be granted and the proceedings moved to a community less affected by the crimes in question. The 1992 trial of Los Angeles police officers in the beating of Rodney King, for example, was moved from Los Angeles County, where the events took place (and were widely covered and inflamed the community), to Ventura County. As another example, having petrified the entire Washington, DC, area in 2002, the snipers John Muhammad and Lee Malvo were tried in southern Virginia.

The objective of changing venue is to minimize pretrial emotional bias. The present model suggests that the effect can be dramatic, but for a subtle reason. For example, in the preceding run, pretrial conditioning occurred and dispositions to convict were rising before the courtroom phase began (see Figure 77). A perfectly effective change of venue would eliminate this pretrial learning, and jury opinions regarding guilt would be based solely on courtroom proceedings. To capture this, we rerun exactly the same example as before, but without the pretrial stimuli. The result of this perfect change of venue is a concluding unanimous verdict of innocent, as shown in the terminal Figure 86 (four below). However, a number of mechanisms are involved.

Recall that, in the prior example, the courtroom evidence (in contrast to the court of public opinion) favored innocence. So, while some probability of guilt is assigned, it is low, as shown in Figure 83.

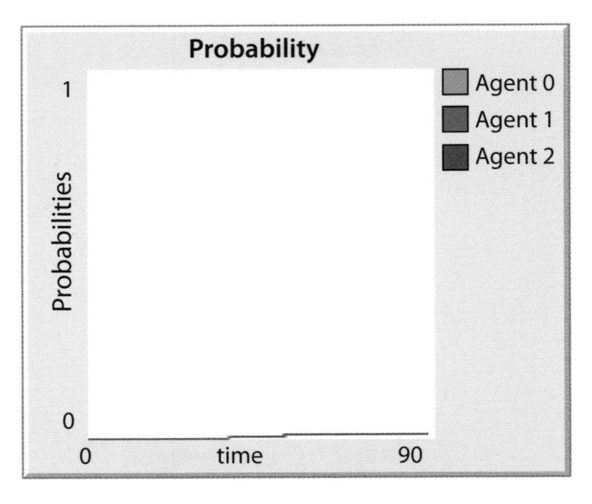

FIGURE 83. Probability of Guilt

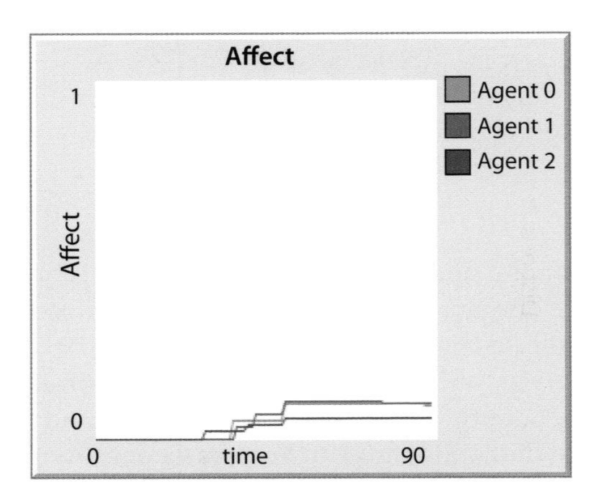

FIGURE 84. Affect with Presentation of Evidence

Juror affects do rise with the presentation of evidence, but the levels are modest, as plotted in Figure 84.

Then the jurors retire to deliberate. Here they interact directly with one another for the first time. Their weights on one another depend on affective homophily, and those with similar feelings are mutually reinforcing. But, as indicated in Figure 85, the levels of affect, and hence the weights, are far lower than before (as also indicated by the lighter red links in the right frame).

This results in *less dispositional amplification* by homophily-based weight growth. In contrast to the previous unanimous guilty verdict, we now see a unanimous disposition to acquit! This is shown in Figure 86.

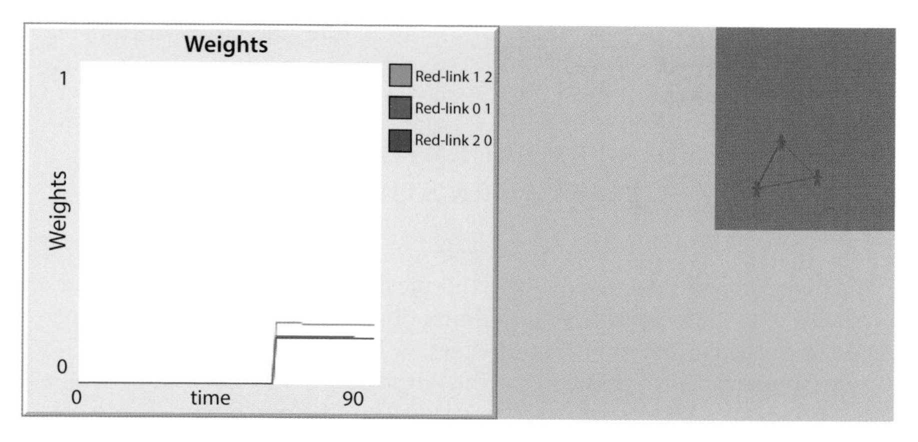

FIGURE 85. Weaker Dispositional Amplification and Lighter Network

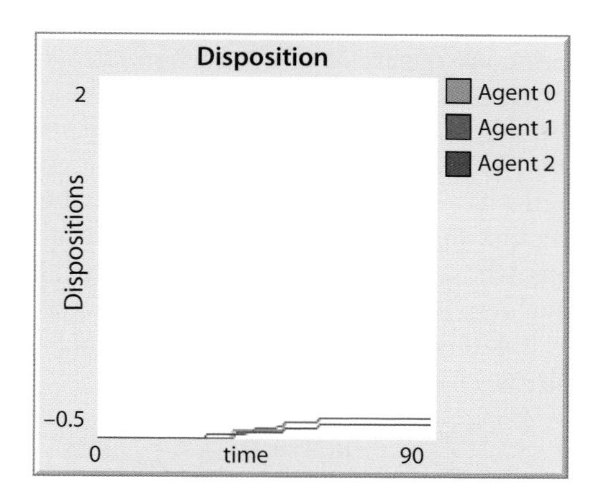

FIGURE 86. Unanimous Disposition to Acquit

Mechanism

Notice that, by our dynamic weight mechanism, the effect of a change of venue extends all the way into the private jury room. How? Pretrial affective growth is eliminated. So, only the affect accumulated in the trial phase proper is carried into the jury chamber. This is lower than it would be without the change of venue. But, because it is lower, *the interagent weights—which depend on both strength and homophily—are also lower*. These made all the difference before! Hence the intrajury momentum and peer effects, so decisive in the former run, are suppressed in this venue, resulting in a verdict of innocent. So, the model suggests that changes of venue may have unappreciated effects, shaping social dynamics *within juries* at times and places far from the scene of the (alleged) crime.[183]

III.10. EMERGENT DYNAMICS OF NETWORK STRUCTURE

One might define the term *network structure* as the continuous vector of interagent weights defined on $[0, 2\lambda]$. Typically, however, the term "network *structure*" is understood in the tradition of graph theory. Structure proper concerns the pattern of internode edges, or links, not their strengths. With some important exceptions,[184] in this literature, links exist or not; it is a binary matter. And the analysis proceeds from there to study various statistics, such as the network's degree distribution, clustering coefficient, and myriad other features.

We, too, want to study the dynamics of structure proper. How then can we map our continuous dynamics of connection *strength*—of interagent weights—onto a dynamic of binary *structure*? One solution explored here is to introduce thresholds and posit that *a link exists if and only if the internode strength exceeds threshold.* As we know from the foregoing, the time series of connection strengths based on affective homophily can be extremely spiky and volatile. But a link threshold filters these spiky dynamics onto binary $\{0,1\}$ step functions, whose value is unity (i.e., a link exists) if the link threshold is exceeded and zero otherwise (no link exists). Mathematically, letting $L(t)$ be the binary link function, and τ_L the link threshold, we simply define the Boolean link between Agents i and j as:

$$L_{ij}(t) = \begin{Bmatrix} 1 \text{ if } \omega_{ij}(t) > \tau_L \\ 0 \text{ otherwise} \end{Bmatrix}. \qquad [58]$$

Equivalently, using the Heaviside unit step function introduced earlier, this is[185]

$$L_{ij}(t) = H(\omega_{ij} - \tau_L). \qquad [59]$$

[183] It is interesting to compare jury trials to elections. The perfect change of venue eliminates Phase 1—the public square phase—of the dynamic. The voting booth removes Phase 3—a nonanonymous voting process. In an election, the analogue of the public phase is the precampaign jockeying and buzz around potential candidates. Then there is the formal phase of official campaigning and formal debate (a term I use advisedly), the "courtroom," where candidates present their respective cases. But then, votes are cast privately in the voting booth. So, the momentum of the jury room itself (Phase 3) is minimized.

[184] Famously, Granovetter (1983) distinguishes between weak and strong ties. These do not change dynamically.

[185] By the symmetry of formula [52], $L_{ij} = L_{ji}$.

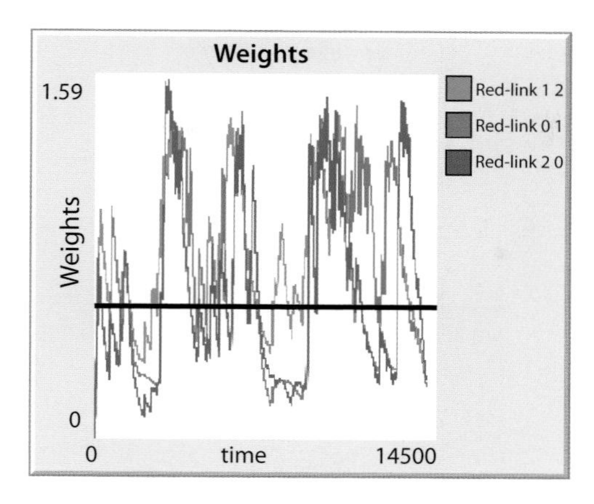

FIGURE 87. Episodes of Connection

This projects the continuous *strength* dynamics on the real interval [0, 2λ] onto a discrete *structure* dynamic on the binary set {0, 1}.[186] Structure thus emerges as a kind of projection, or filtered version, of underlying connection strength dynamics. So, the model crudely addresses the questions: Why does network structure happen? And why might it change?

As an example, in the time series of Figure 87, we define weights as in equation [52]. Weights vary between 0 and 2. If the link threshold (black) is set to some intermediate value, we see that links form and dissolve as the undulating affinity landscape rises above and falls below the threshold. There are some fully connected periods (where all weights exceed threshold), some completely disconnected episodes (where none do), and some episodes in which only two of the three nodes are connected, as where only the green curve (the link between Agent 0 and Agent 1) exceeds threshold.

Networks "happen," in other words, when weights "break the surface" defined by these thresholds, as suggested in Figure 88, with a blue subsurface.

Network Structure Dynamics as a Poincaré Map

Thus we see that network *structure* as classically defined (binary) is an epiphenomenon of underlying similarity dynamics (continuous). It is a *suspension*,[187] somewhat analogous to a Poincaré map. The threshold is the Poincaré section. These breachings of the surface (the threshold) generate

[186] If one wanted an everywhere differentiable version, one could define L as a sigmoid of ω_{ij} confined to the range [0,1] and tune the function to differentiably approximate the step function as closely as desired.

[187] See Jackson (1992).

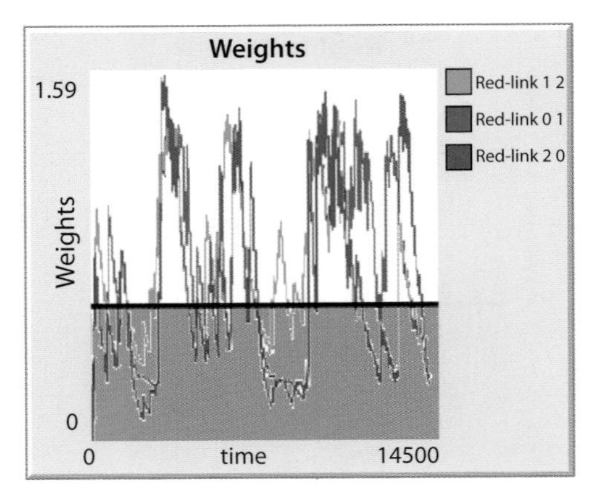

FIGURE 88. Networks with Threshold

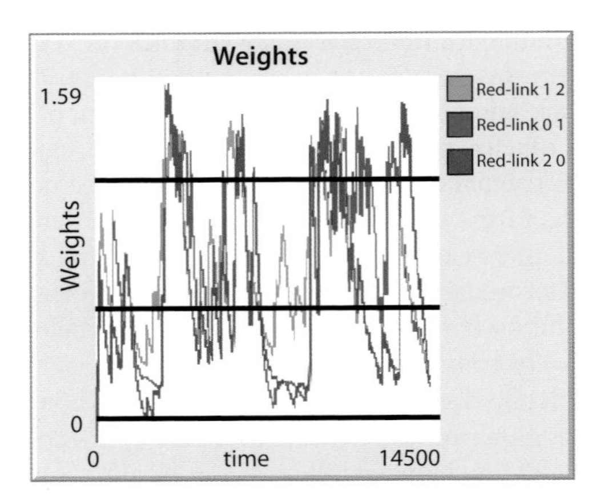

FIGURE 89. Network Dynamics on Affective Similarity

the discrete time return map of an underlying continuous dynamic. [On Poincaré maps, see Guckenheimer and Holmes (1983), Wiggins (1990), Strogatz (2001), and J. M. Epstein (1997)]. In this view, then, *dynamic network structure is the binary (step-functional) projection of an underlying continuum affinity phenomenon.*

Obviously, the binary dynamics of network structure change if we change the link threshold (the Poincaré section). Figure 89 shows the surface just depicted as the middle threshold, but with a higher and lower one shown as well. Given the lowest threshold, the network is fully connected throughout. As all this is happening, the degree distribution of the network and innumerable other measures of connectivity are also changing.

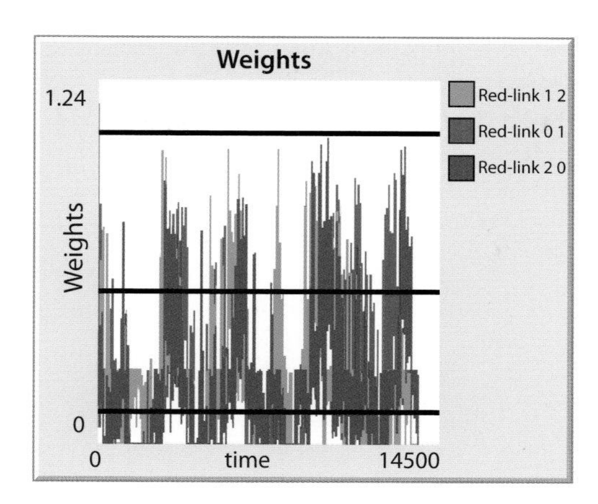

Figure 90. Network Dynamics on Cognitive (Probability) Similarity

This is network dynamics on *affective* (V) similarity. But using the same algebraic form of equation [52] with P instead of V, one could study network dynamics on cognitive similarity—that is, on the similarity of probability estimates on the same landscape. That picture is given in Figure 90, also with three thresholds.

However, unlike the high-threshold dynamic on affect, the high threshold (upper black line) here produces disconnection at essentially all times. Comparing the low threshold cases, on affect full connection obtains; while on probability, there are many (and well-packed) periods of complete disconnection. This comparison suggests what is, of course, true: some of our affect-based networks (dog lovers) may be thriving while deliberation-based networks (chess clubs) are in decline.

Many other network visualizations are possible. Obviously, one can literally draw links forming and dissolving as agents interact spatially. Thinking again of affective homophily, if we posit a link threshold of 0, full connection obtains; while at a threshold of 1, connection is extremely sparse. **Movie 12** shows an animation of connection dynamics at an intermediate threshold of 0.5. Here connections form and break in the course of the run. (All code, once again, is provided on the Princeton University Press Website). The upper-left frame of Figure 91 shows the agents on a completely quiescent landscape. Since there is no stimulation, affect remains 0, so connection weights are also 0. Connectivity changes as agents move into and out of stimulus zones, as shown in the other three frames, captured at arbitrary times. Of course, weights never decline if there is no affective extinction rate. Here, we assume a positive extinction rate.[188] In yellow zones, there

[188] See the Appendix IV table of numerical values.

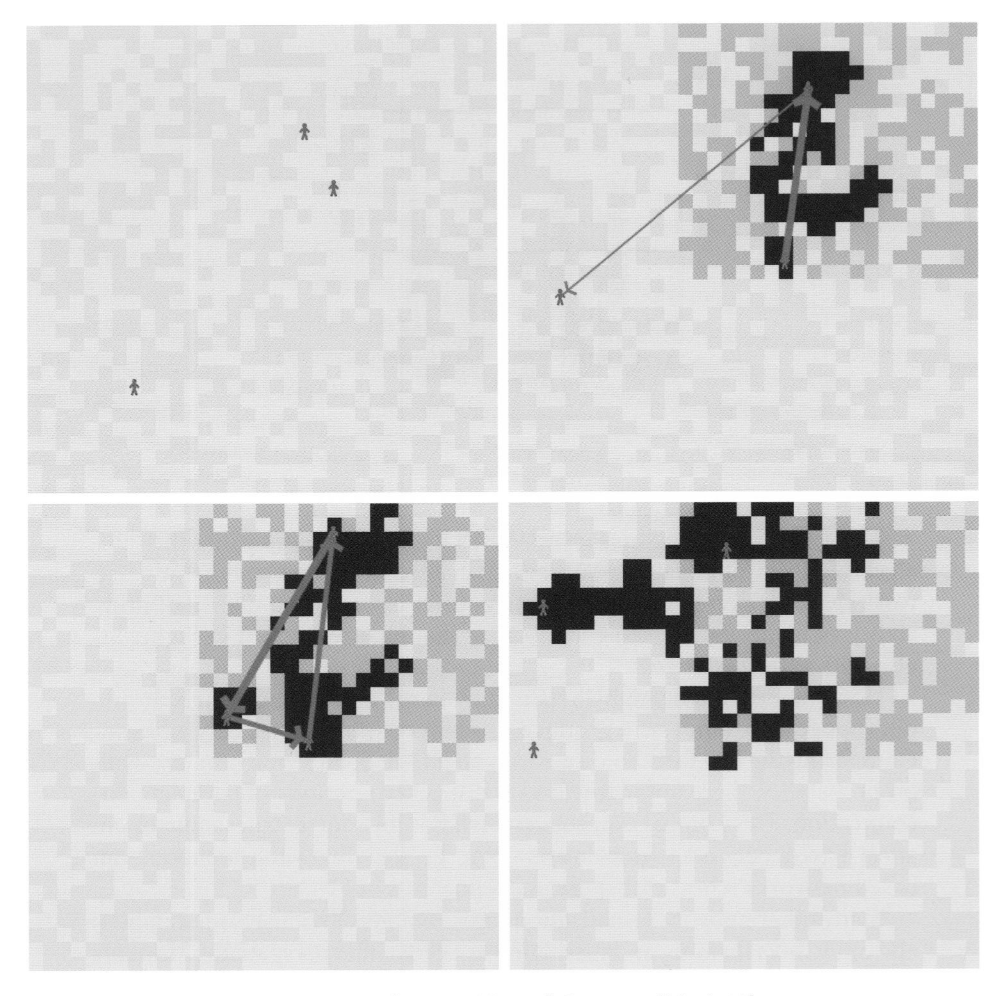

Figure 91. Endogenous Network Structure. [Movie 12]

is no stimulus, so affective extinction is unchecked. This can lead to link weakening and breakage, as in the right frames.

Another visualization of structural dynamics is to display the derived step functions themselves. We illustrate this in stages.

First, for arbitrary parameters, we give time series of raw affect itself (Figure 92).

Second, in Figure 93, we plot the weights on affective similarity (as in equation [52]) with no binary filter—that is, no *L*-function introducing a link threshold and mapping values to $\{0, 1\}$.

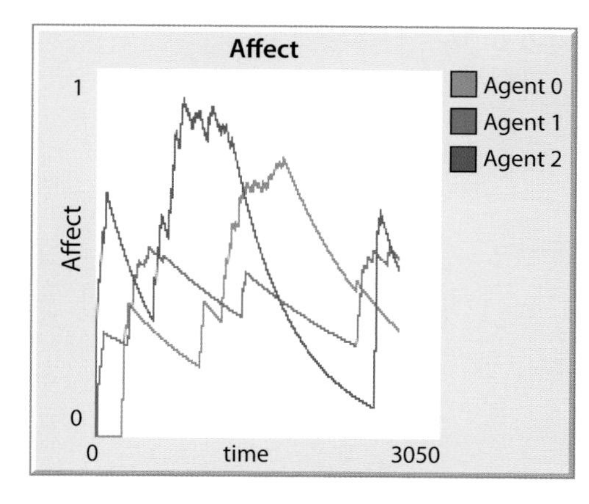

FIGURE 92. Raw Affect

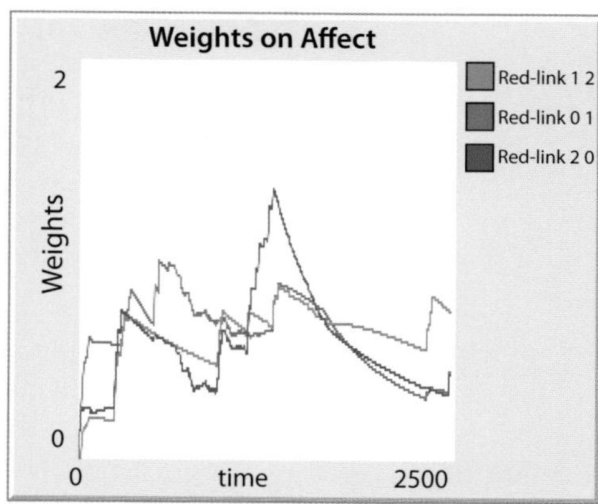

FIGURE 93. Dynamic Weights

The next three figures, Figures 94–96, give this weight plot, filtered to produce different dynamic network-structure trajectories. The first sets the link threshold to 0.25, meaning a link is established if and only if the weight (ω_{ij}) exceeds 0.25. Since this is relatively weak, the network quickly becomes fully connected and stays that way. Note that a link exists if the value is 1 and does not exist if it is 0. I include the range −1 to 2 rather than only the operative range, 0 to 1, merely for visual clarity. All the action is on {0, 1}.

Link Threshold 0.25

Periods of full connection become rarer as we hike the link threshold to 0.5.

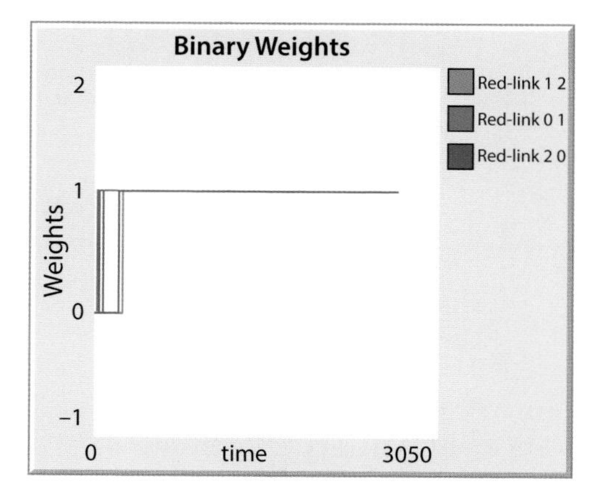

FIGURE 94. Threshold 0.25

Link Threshold of 0.5

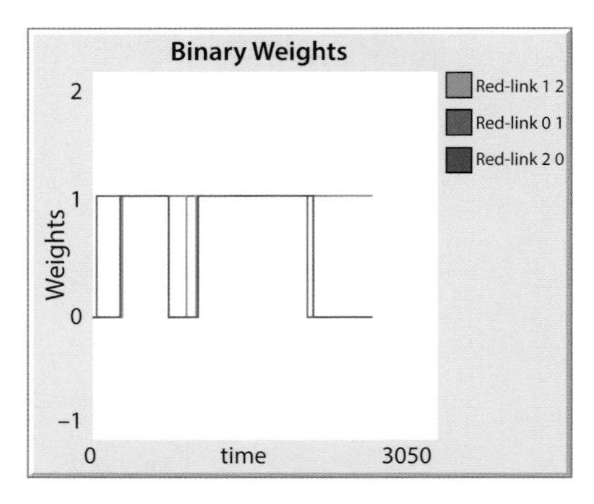

FIGURE 95. Threshold 0.5

Finally, disconnection dominates at threshold 1.0, with no periods of full connection, as shown in Figure 96.

Link Threshold 1.0

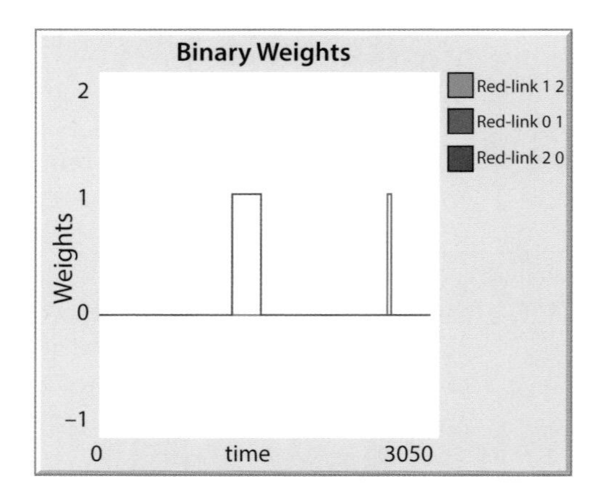

FIGURE 96. Threshold 1.0

Network Structure an Epiphenomenon of Affinity Dynamics

So, in this conceptualization, underlying affective dynamics (grounded in a classic learning model on a dynamic landscape of stimuli) generate time series of interagent weights (based on a strength-scaled homophily measure). But this same weight dynamic generates network structure proper (the pattern of binary edges and nodes) when it is mapped onto a binary step function through a link threshold. *Each different link threshold generates a different dynamic network structure for the same affective affinity evolution*, just as a different Poincaré section may generate a different return map for a fixed dynamical system.[189] Agent *behavior* (to rebel or not) depends on total disposition minus the action threshold. As we have shown, each dispositional trajectory (and hence action history) is compatible with many different dynamics of network *structure*, defined narrowly as the set of binary internode connections. Every choice of a link threshold selects one of these dynamic evolutions of network structure proper. In the Arab Spring example, if one posits a threshold of zero, then full connection occurred very early in the affective dynamics.

Relation to Literature

The preceding model of *attachment by strength-scaled homophily* (whether affective or probabilistic) offers a completely different network formation mechanism than *preferential attachment* (PA). There, the probability of a new node attaching (making an edge) to a node k is proportional to the latter's degree—the number of nodes to which k is already connected. Thus, well-connected nodes enjoy an (increasing) advantage in collecting new partners. The idea has a long mathematical lineage.[190] In 1976, de Solla Price introduced a general cumulative advantage distribution to explain dynamics in which "success seems to breed success."[191] In 1999, Barabasi and Albert (1999) presented a model in which PA was shown to generate scale-free power-law degree distributions observed in many large networks, including the World Wide Web.

It may well be that PA has been a central mechanism in the formation of the World Wide Web. But, in many contexts, and in my preceding model, attachment is unrelated to degree, depending instead on an associative dynamic.

Degree-Independent Attachment Models and an Hypothesis

In research on the coevolution of networks and political attitudes, Lazar (2001) uses the absolute value of an attitudinal (though not literally affective)

[189] The converse does not obtain. Different dynamical systems could generate the same return map, given a fixed Poincaré section. The same holds for binary network structure in this model. If one doubles all amplitudes with frequency fixed, the continuous dynamic is different, but its binary projection is not.

[190] See Simon (1955) and Yule (1925).

[191] See Price (1976).

difference as his measure of political similarity, and establishes empirically that "homophily in political views contributes to tie formation."[192] In a different study, Levitan and Wisser (2009) demonstrate a strong correlation between attitude *strength* and social ties. This relationship, like Lazar's, is unrelated to degree. But unlike Lazar, they ignore homophily. My model includes both a strength term (e.g., the sum of affects) and a homophily term. I certainly would not claim that either of these studies directly tests my model (or even defines terms in exactly the same way). But the model does offer a testable hypothesis, namely, that ceteris paribus, *the tie probability increases with the product of affective strength* $(v_i + v_j)$ *and affective homophily*, $1 - |v_i - v_j|$.

It would, of course, be interesting to see whether a scaled-up version of the model could generate the observed dynamics of real-world networks or simply generate known network structures, such as are hypothesized for obesity, smoking, and other behaviors.[193]

My exposition here has been predominantly on affective strength and homophily. But it could be on similarity of probability measures, or dispositions, or spatial proximity, as well. Again, for various link thresholds, these would generate a diversity of network structure dynamics that could be compared to data.

Albeit crudely, this captures a central fact about people. We belong to many networks at any one time. We can belong to a network based on data and empiricism, $P(t)$, and an entirely different network based purely on emotional, $V(t)$, similarity. They are both dynamic and may overlap at various times, or not. To expand on Spinoza, then: we are not just social animals. We are many social animals at once!

While many of the earlier computational parables have been dark, the parable of the Arab Spring shows that, unlike economics (discussed further shortly), the science of *Agent_Zero* is not necessarily dismal. Lynchings and liberations are both generable.

III.11. MULTIPLE SOCIAL LEVELS

Now, all examples thus far involve the activation of the yellow spatial sites. Call these the Level_1 activations. Then Level_2 agents take these as their Rescorla-Wagner conditioning trials and take actions accordingly. Thus far, only those two layers have been in play. But there is no reason to stop there.

[192] See Lazer (2001) and Carpenter, Lazer, and Esterling (2003).

[193] See Christakis and Fowler (2007) and the methodological literature surrounding their work, beginning perhaps with Lyons (2011) and Shalizi and Thomas (2011).

One could posit Level_3 agents who take the deeds of Level_2 agents as their stimuli. They train on those stimuli by the same general Rescorla-Wagner scheme and also act accordingly. Presumably, their behavioral repertoire—their possible actions—would be qualitatively different from those of Level_2 agents (e.g., they might have the option of regulating Level_2 agents). This process of adding levels can, of course, be continued. In general, the deeds of Level_*n* agents are the stimuli for Level_$(n + 1)$ agents. A concrete example for two levels follows.

Agent_Zero *as Witness to History*

Recall our very first parable, the slaughter of innocents. For the *perpetrators* of slaughter, the stimuli (the Rescorla-Wagner training trials) were the orange activations on the landscape, interpreted as attacks by insurgents. When the perpetrator's threshold was exceeded, he indiscriminately wiped out all patches—orange *or* yellow, guilty *or* innocent—within some destructive radius, coloring them a dark blood red. That was the end of it. No agents ever came along and took the dark red patches as their stimulus. Let us introduce such an agent—she could be a UN peacekeeper. We color her green, as shown in Figure 97.

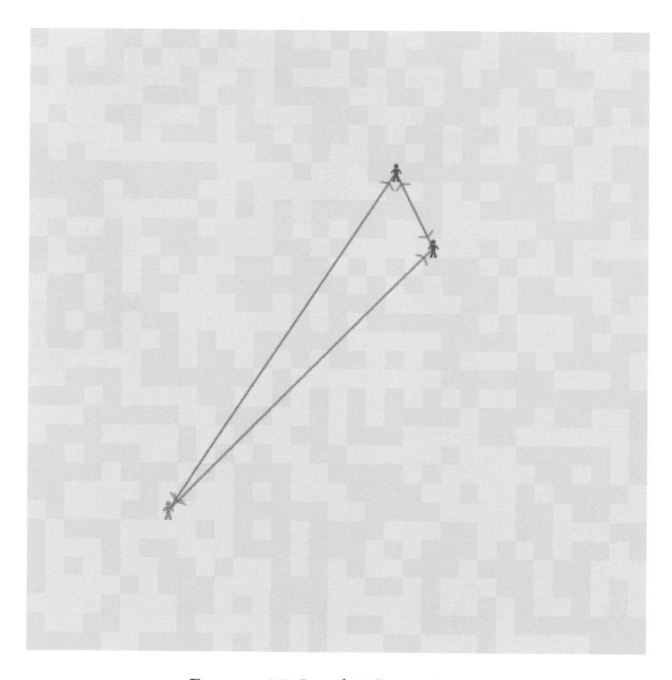

FIGURE 97. Level 2 Green Agent

For this agent, it is not the orange patches, but rather the blood red ones (the dead, in other words) that are her training trials. Activations on the landscape (Level_1) stimulate the Blue perpetrators (Level_2); their deeds stimulate our Level_3 Green agent. She roves the landscape like the two Blue perpetrator agents. She may come upon killing fields in the course of her random walk. She updates her affect exactly like all other agents—according to the Rescorla-Wagner scheme—but the other agents' victims (the dark reds) are her stimuli, not the orange patches that stimulate the killers. Every time she sees a red patch (a dead agent), she updates her affect. We record this event by having her paint the patch grey. (You may imagine that she erects a grey tombstone or, more likely, files a grey report).

Illustrative Run

In this run, it takes her (a Level_3 agent) a while to discover the terrible truth about what is happening at Level_1. In this particular random walk, she hasn't encountered any killing fields at the point shown in Figure 98.

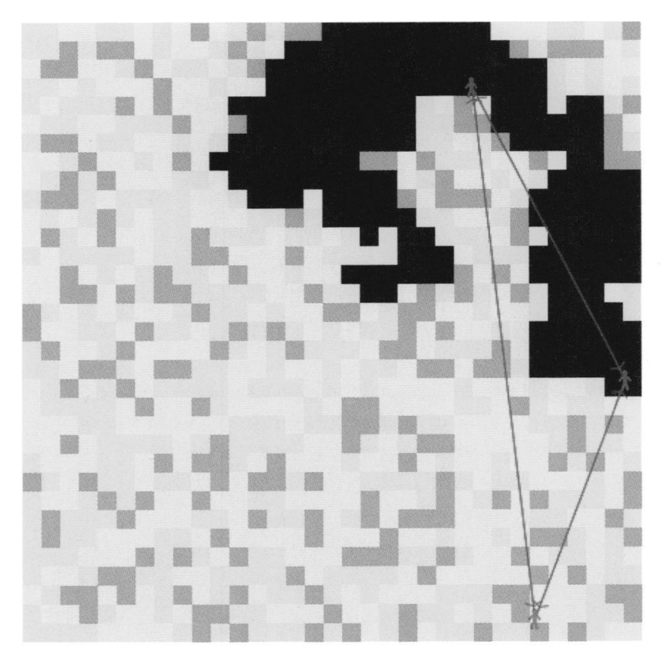

FIGURE 98. Killing Fields Not Yet Discovered (links for visualization only)

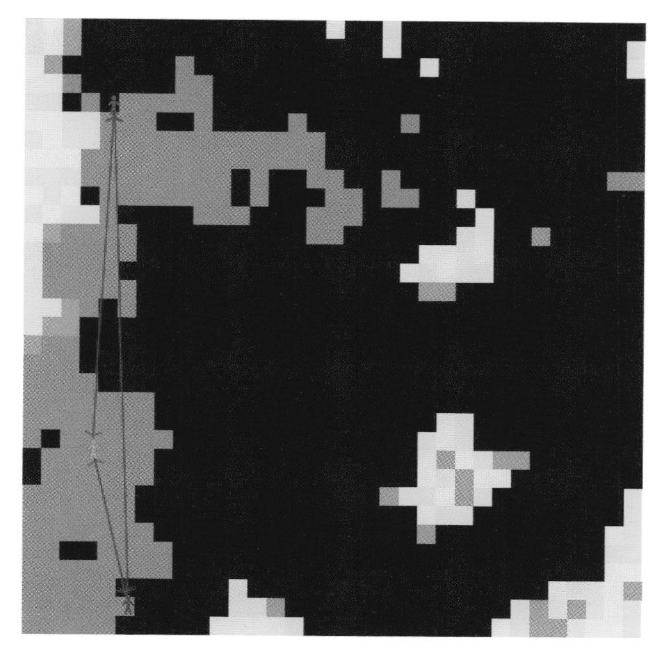

Figure 99. Killing Fields Marked

But eventually she does. Soon, she encounters killing fields, begins updating her affect, and begins marking the killing sites (Figure 99), perhaps to document events for the Level_4 International Court of Justice in The Hague.

Even at this stage, her (Green) Rescorla-Wagner acquisition curve is flattening out (Figure 100). Her associative capacity is approaching its limit, and each new discovery is less impactful than the last (diminishing marginal impact is evident, in other words), even as she paints the world grey (Figure 101).[194]

This, of course, is but one multilevel interpretation. The dark red patches could be vaccines refused and the Green agent a public health authority observing instances of this behavior—and becoming energized (through Rescorla-Wagner updating) to intervene and change it. Or the dark red patches could be assets abandoned in a financial panic, and the green agent the Federal Reserve growing ever more fearful of deep economic repercussions.

[194] This brings to mind the important phenomenon of "psychic numbing" documented in Slovic (2007).

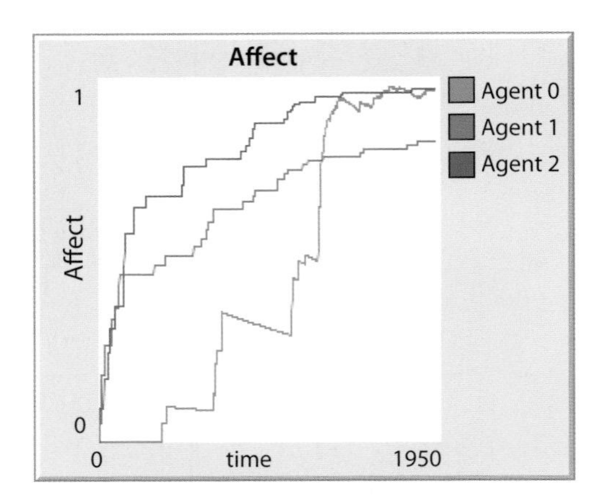

Figure 100. Affective Dynamics

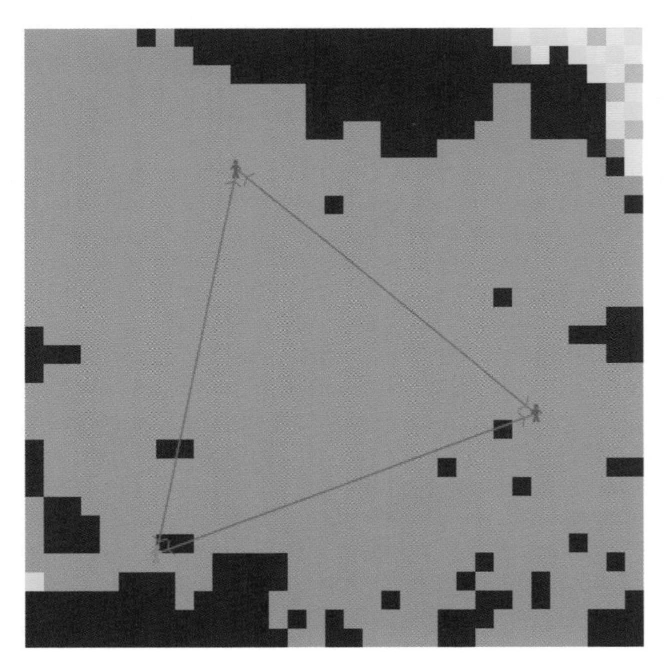

Figure 101. Holocaust Revealed

The Recursive Society

The point is that, in the main exposition, we have not gone beyond the dark red patches (jasmine for the Arab Spring). But they can themselves serve as the stimuli for a higher layer of agents, whose actions (grey patches) can serve as stimuli for a yet higher layer, and so on. We can add layers recursively in this way to build up a multilayered model in addition to the two-layered (patches and rover agents) one that has been our focus. Within each layer, moreover, social dynamics can unfold through weights and homophily, by the same apparatus we've used at the lowest agent level. As emphasized earlier, the Rescorla-Wagner model applies not only to fear but to other emotional learning. So, one layer could be growing more fearful through its learning dynamic, while a different layer could be growing more happy.[195] All in all, a society of tremendous depth and richness can be generated in this recursive fashion, with different network, action, and emotional extinction dynamics unfolding at the different levels. One could even allow Level_*n* agents to reach down and act directly on agents far below them in the hierarchy. All this strikes me as an important line of future work.

III.12. *THE 18TH BRUMAIRE OF* AGENT_ZERO

To this point, agents have not had offspring. Many agent-based models include some form of reproduction. Some have also explored the intergenerational transmission of genes, cultures, and immune systems, idealized in various ways (Epstein and Axtell, 1996; Miller and Page, 2007, Ch. 10). Here, let us attempt something different, an intergenerational parable inspired by Marx. In *The Eighteenth Brumaire of Louis Bonaparte* (1869; 1972 ed.), he writes,

> Men make their own history, but they do not make it just as they please; they do not make it under circumstances chosen by themselves, but under circumstances directly encountered, given, and transmitted from the past.
>
> Then, inimitably, he writes,
>
> *The tradition of all the dead generations weighs like a nightmare on the brain of the living.*

[195]I thank Julia Chelen for this thought.

How might we interpret and generate this parable in *Agent_Zero*? The first (nonitalicized) passage requires that agents be born into preexisting circumstances. This we can represent by having them begin life in the immediate (von Neumann) neighborhood of their parent, ensuring that their "directly encountered" environment is, as per Marx, not of their own choosing. His (italicized) continuation, however, is the enticing and more elusive part, the idea that parents transmit a "tradition" that "weighs" on their offspring.

Transmission of Chronicles

How might we get at this? As a crude beginning, recall that, in each period agents are making a social appraisal, "calculating," as it were, the relative frequency of enemies (orange) within their vision. Employing the memory apparatus developed earlier, they carry a string of the *m* most recent such appraisals. We have called *m* the memory length and the string that agent's memory. Parents do, in some sense, impart memories to their children—chronicles of their culture's achievements, its persecutions, its days that will live in infamy. The parent may be gone, but the parent's story, her chronicle, her "tradition," if you will, lives on in the offspring. While we inherit our parents' chronicles, we also leave home, have new experiences, and accumulate a new string of social appraisals. The parental nightmare—distrust of yellow sites, for instance—can be overwritten by happier stories, or—to be sure—by new nightmares! We will tell the happier parable in code.

To wit, we imagine the world as divided into a hostile north and a peaceful south. The parent lives her life in the violent north (as in the upper left panel of Figure 102). Accordingly, her narrative (her period-10 memory string) consists entirely of high values—appraisals of high likelihood that any agent within vision is an attacker, an enemy.

At a randomly chosen time, she clones an offspring. The parent passes an exact copy of her memory—her experiential chronicle—to the child. And then she expires (her grave is marked by an X in the right panel), becoming Marx's dead generation. So, at this moment, the "tradition of the dead generation weighs like a nightmare on the brain" of our newborn. But the child begins to move away and begins to overwrite the inherited chronicle with his own. Eventually, after some time in the violent north, the child migrates to the sunny south (Figure 103), where his experiences are all peaceful (run the Applet on the book's Princeton University Press Website to generate this spatial history).

His social appraisals eventually overwrite the transmitted story of yellow hostility until he is liberated from it. The dead generation's nightmare is lifted.

The story of this awakening is recounted in his memory strings, which begin with a duplicate of the parent's nightmare. But through migration

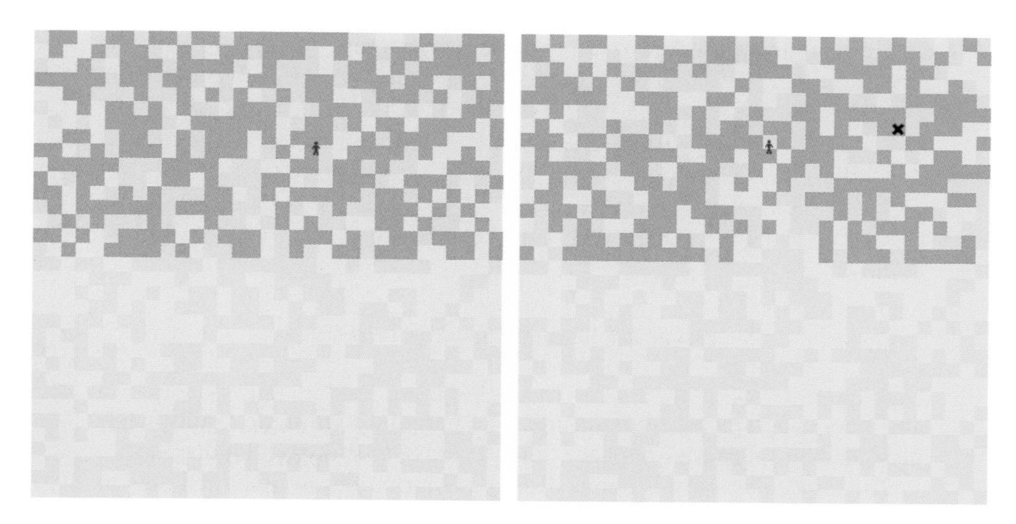

Figure 102. Birth and Death

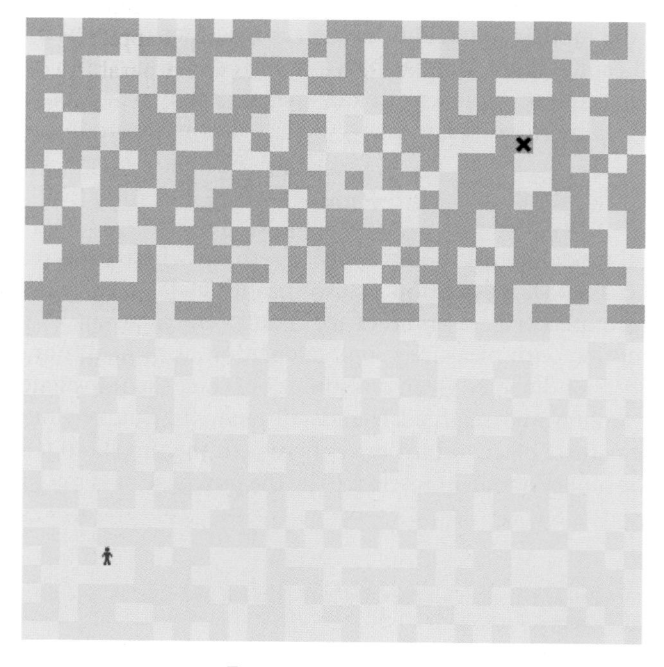

Figure 103. Migration

Age	Memory
0	[7, 6, 6, 6, 6, 5, 5, 5, 6, 7]
10	[5, 5, 5, 4, 2, 2, 1, 2, 0, 0]
15	[5, 4, 2, 2, 1, 1, 0, 0, 0, 0]
20	[2, 1, 1, 0, 0, 0, 0, 0, 0, 0]
25	[0, 0, 0, 0, 0, 0, 0, 0, 0, 0]

FIGURE 104. Inherited Chronicle Overwritten

and new experiences, zeros invade. The parental nightmare is ultimately displaced by a narrative of peace (all red zeros), as shown in Figure 104.

The particular story is not as dismal as Marx's. But we will turn now to what Carlyle dubbed "the dismal science," economics.

III.13. INTRODUCTION OF PRICES AND SEASONAL ECONOMIC CYCLES

In very general terms, we have discussed a number of economic interpretations of the model: contagious financial panic, fear-inspired dumping of assets, and capital flight. But, we have not introduced prices in any explicit way. And, to economists, prices are the sine qua non of true economic models. But, prices are so easily introduced that a brief discussion is warranted.[196]

Prices

The general model, recall, would apply to settings where decisions depend on affect (passion), calculation (reason), and social network effects. In choosing products (e.g., a brand of ice cream), the associative strength of its consumption with pleasure v, our calculation of its health implications (probability P of diabetes given consumption), and the prevalence of consumption among my weighted friends all enter in. But, obviously, so do *prices*. The natural place to introduce them is in the threshold term. Let us imagine the binary action of interest to be the purchase of some good[197] and

[196] Again, detailed assumptions are provided in the table of Appendix IV, and the Source Code contained in the Applet for this run.

[197] To accord with our binary action model, one can assume this good to be *indivisible*, in the terminology of microeconomics.

make the threshold a nice orthodox convex function of exogenous price (p). Specifically, set

$$\begin{aligned} &\tau = \tau(p) \text{ where} \\ &\tau'(p) > 0, \\ &\tau''(p) < 0, \\ &\tau(0) = \tau_0. \end{aligned} \qquad [60]$$

All else fixed, consumption decreases as price increases, because the threshold rises with price, but it does so at a diminishing marginal rate.

The economist, Wesley Mitchell (1927, p. 236) wrote that, "Two types of seasons produce annual recurring variations in economic activity—those which are due to climates and those which are due to conventions."

As a trivial conceptual demonstration of the former, consider the demand in New Zealand for fresh fruits and vegetables—apples and kiwi fruit, for example. Local production stops for winter. So, the price increases to cover transportation costs from elsewhere. Seasonal fluctuations in price result, as shown in Figures 105 and 106.[198]

Ceteris paribus,[199] demand and price move in opposite directions, so these price fluctuations generate anticorrelated demand dynamics. Thinking of disposition as the disposition to purchase at prevailing prices (so, as in neoclassical perfect competition, *Agent_Zero* is a price taker), we can easily mimic this cyclical demand dynamic by forcing the threshold seasonally. As an off-the-shelf function, I have let the threshold vary as the cosine of elapsed time, as is often done in modeling seasonal epidemics (J. M. Epstein et al., 2007; Earn, Rohani, Bolker, and Grenfell, 2000; Koelle, Pascual, and Yunus, 2005; Ellner et al., 1998; Aron and Schwartz, 1984). This produces the inverse "price-demand" cycles shown in Figure 107.[200]

Prices and demand are negatively correlated and cycle periodically, as in the fresh-fruit examples.

Exogenous Climatic Variation

In this example, seasonally varying *production costs* are driving the price fluctuations. It is more expensive to produce, refrigerate, and transport this commodity, fruit, from Florida to New England in winter than to produce and consume it locally in New England in the summer. And these increased winter costs are passed on to consumers as a higher price in New England.

[198] Seasonal Price Fluctuations for Fresh Fruit and Vegetables (n.d.).

[199] We ignore Giffen goods and assume strictly negative price elasticity of demand, for example.

[200] All agents have the same threshold, and 2 is drawn last, so the curve is blue.

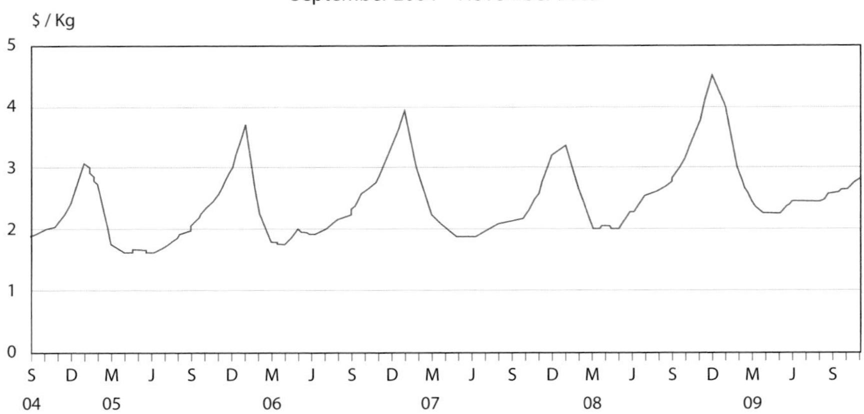

Figure 105. Apple Prices. Source: Statistics New Zealand Tatauranga Aotearoa

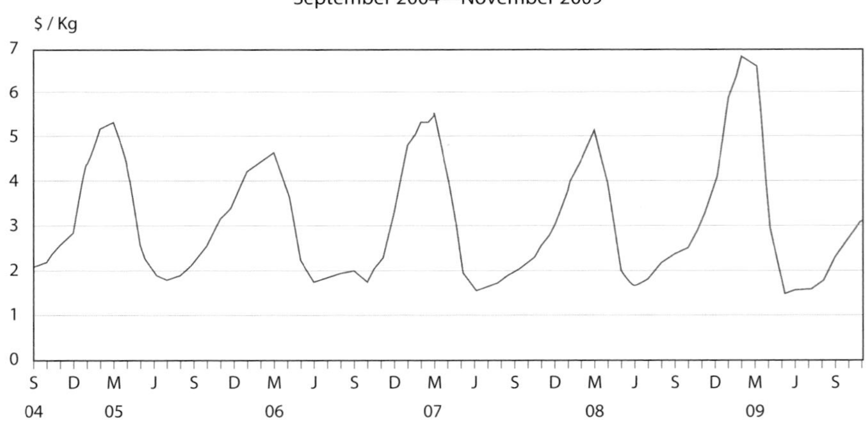

Figure 106. Kiwifruit Prices. Source: Statistics New Zealand Tatauranga Aotearoa

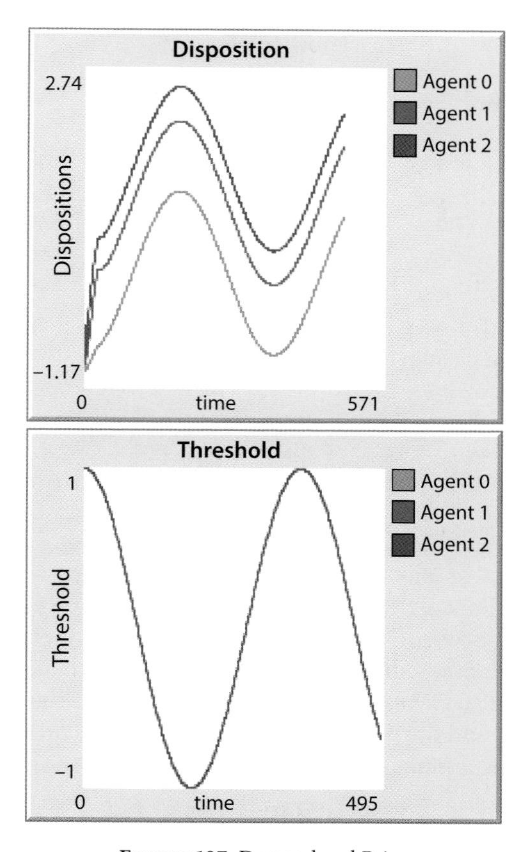

FIGURE 107. Demand and Price

In the *argot* of textbook economics, the supply curve shifts left raising (equilibrium) prices at each quantity of output.[201]

[201] A textbook economic supply curve rises to the right. The neoclassical parable here is that, as prices increase, the prospect of profit stimulates new firms to enter the market. The horizontal sum of outputs at which each producer maximizes profit (the output quantities at which each firm's marginal production cost equals the exogenous market price) constitutes the market supply at that price. If the exogenous price falls, the entire supply curve shifts to the left, by the same reasoning. In perfect competition, there are no barriers to entry, and no producer can affect price. Firms are "price takers." But this leaves no way to account for price dynamics. After all, if every producer is a price taker, who changes prices?

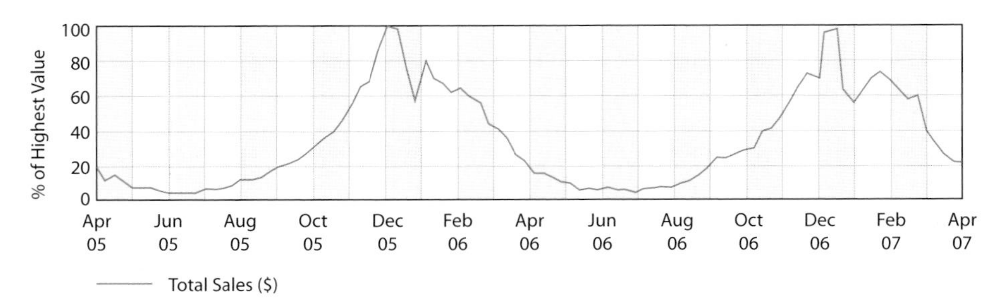

FIGURE 108. Demand for Sporting Goods Spikes at Christmas: eBay Category Trends (skiing and snowboarding).

Endogenous Conventional Variation

However, changes in demand (rather than supply) can occur due to seasonal fluctuations in affective dynamics, holidays being a prime example; these are Mitchell's "conventions." Skates and snowboards presumably cost no more to produce in December than in June. But their prices spike at Christmas. In this case, affect and the disposition to purchase are seasonally forced. There is also clear extinction of affect after the stimulus of the holiday is turned off, and the price data show a decrease accordingly. See Figure 108, depicting the annual spikes in Christmas sales of sporting goods.

Seasonal Demand

How shall we generate this in the *Agent_Zero* model?[202] We will exercise the model in yet another way, allowing the landscape to oscillate seasonally. But we now interpret our space as the shopping district of a major city. Yellow patches are products of any sort. Orange patches are products that are Christmas specific and/or bear Christmas-specific advertising. In the former (Christmas-specific) category would fall Christmas trees (real or artificial), decorations, lights, electric plastic Santas and reindeer for your roof, and Christmas carol CDs. Products that are not Christmas specific but

[202] There are (at least) two ways of doing this in the equation-based versions of the model. One is to seasonally force the affective differential equations, setting: $dv/dt = \alpha\beta(\lambda - v)k\sin(\xi t)$, for example. Another approach would be to apply a seasonal forcing to the *solution v(t) of the unforced* differential equations, when composing the Disposition. One could also force the probability seasonally. The spatial forcing adopted affects V and P at once, and exercises the model in a new way. I thank Julia Chelen for invaluable consultation on this.

bear Christmas-specific advertising could include skates, bikes, sleds, and holiday hams,[203] seasonally festooned with red and green ribbons.

These orange activations are the stimuli associated with the season, and they increase v.[204] The same process also increases P, the local relative frequency of orange. We interpret this as the probability that, when you walk into a mall, for example, you will encounter a Christmas purchase candidate—either Christmas specific or strongly Christmas associated through advertising. As usual, v and P combine to produce a purchase disposition that is compared to a threshold (that can be price dependent, as above). The dispositions of other people also have weight and are detectable by their holiday clothing, caroling, gift-laden shopping bags, and text messages and cellphone calls about good deals just around the corner (which we would thus include in the agent's "vision," or shopping radius). These are the weights, ω_{ji}. As per the familiar skeletal equations, all this is added up and compared to the individual's action (purchase) threshold. If total disposition exceeds τ (or $\tau(p)$ as before), the shopper takes binary action, buying all items in some radius. If the radius is 0, she buys only the orange product. If the radius is 1, she might buy a Christmas-specific sweater (orange patch) but also a generic Burberry's scarf and hat (yellow patches). The shelves where these items were stocked are colored dark red, indicating "sold." Stimuli and dispositions drop sharply after the holiday, the Christmas tree lights go back in the attic, and in the off-season the shelves are restocked and the cycle repeats again the next year. We generate this "Christmas tale" next.

A Christmas Story

Before Christmas season we see our three erstwhile shoppers walking randomly around the Upper East Side against a background of yellow off-season shopping opportunities (left panel, Figure 109). But as the Christmas season begins, the landscape lights up with seasonal offerings, shown in orange (middle panel, Figure 109).

And before too long, affect and probability drive dispositions over the shoppers' (heterogeneous) thresholds and the shopping sprees begin, leaving a trail of empty shelves (dark red) where our agents have made purchases, as shown in the rightmost frame. Christmas offerings and other seasonal stimuli then end, the landscape reverts to yellow (as on the left), the probability

[203] The unconditioned stimulus (US) would be the innate gustatory enjoyment and sustenance provided by ham. The conditioned stimulus (CS) would be the Christmas-specific presentation thereof: green and red ribbons and other seasonal adornments, placement in decorated areas of the store, and so forth.

[204] Ambient seasonal stimuli—ubiquitous Christmas music and street decorations, Santas on every corner, and the smell of roasting chestnuts—could all increase v_0.

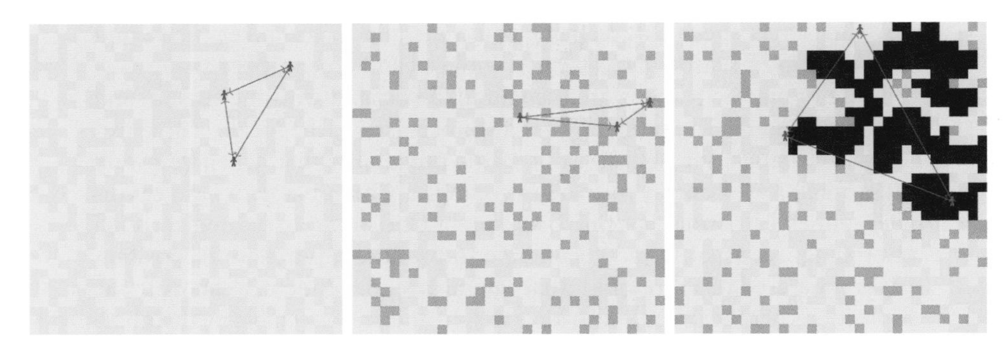

FIGURE 109. A Christmas Cycle [Movie 13]

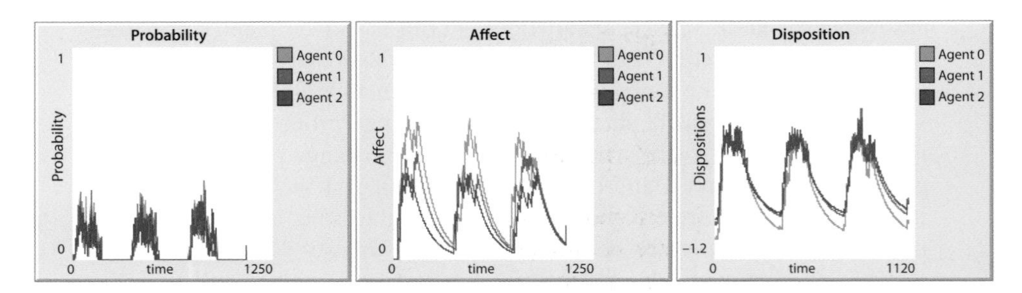

FIGURE 110. Deliberation, Emotion, and Disposition to Purchase over Three Christmas Seasons

of finding a Santa suit for sale drops precipitously, and the holiday affect slowly wanes (extinction)—and with it, the Christmas-specific buying disposition. The cycle repeats on an annual basis, as shown in the three panels in Figure 110, giving probability, affect, and net purchase disposition for our three (heterogeneous) agents over three roughly 365-day years.

In **Movie 13,** posted on the book's Princeton University Press Website, you will see different spatial patterns of available products and buying each season.

Aggregate demand (the sum of all net dispositions) over two years is shown, in Figure 111, to very crudely mimic the previous two-year sporting goods sales data.

Given sufficient action radii, it may be that very little Christmas-specific advertising is required to "tip" the system into seasonal consumption spikes. The framework allows exploration of this possibility.[205]

[205] I thank Michael Makowsky for this suggestion. If it is, in fact, true that advertising by a few can tip the system into a shopping frenzy, then whether or not to advertise might be cast as a Prisoners Dilemma.

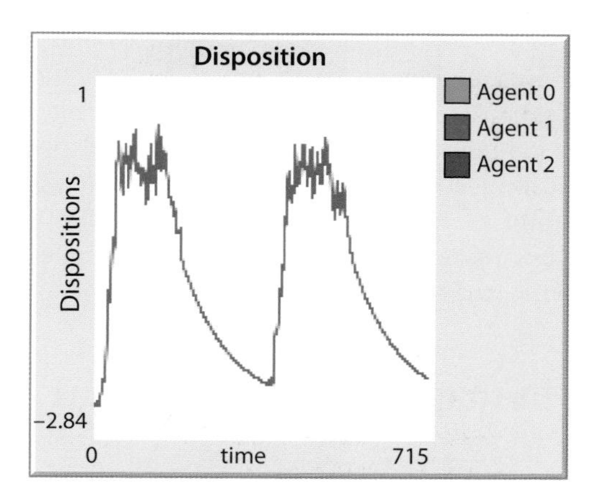

FIGURE 111. Cyclical Aggregate Demand

Serial Dispositional Satisficing

One could extend this by endowing each shopper with an initial budget, B_0, which simply depletes as purchases are made. At any time, the agent's *net budget* is simply the amount remaining. *Net disposition* to purchase an item at a given price, p, is $D - \tau(p)$, as before. Purchase opportunities are encountered at random as the agent moves about the shopping-scape, and purchases are made serially whenever the following simple satisficing[206] condition is met: *net disposition and net postpurchase budget are both positive.*[207] One could even complete this bare-bones sketch by modeling patches as firms owned and operated by *Agent_Zero*–type CEOs, who *set* prices (p) to cover costs and reap profits.[208]

Neuromarketing

Beyond holiday cycles, one can see how a neuroeconomically sophisticated producer would seek to maximize *Agent_Zero*'s disposition to consume her product. The model quite naturally reflects that producers can focus their advertising on *affect* ("Campbell's soup is *mmmmm good*") on health *deliberation*

[206] "Satisficing," like "bounded rationality," is Herbert Simon's term.

[207] Unlike most consumers, our agents won't go into debt.

[208] For example, based on emotions (gut feelings, V, about the market), data (P), and the influence (ω's) of other executives, an *Agent_Zero* CEO might end up negatively disposed toward advertising (idealized as a binary matter). Her products would then never turn orange. But she might still profit, as many yellow products are consumed in the general holiday surge, a kind of externality.

(Healthy Choice soup is low fat), on *social appeal* ("Nobody doesn't like Sara Lee"), on *prices* ("Buy one, get one free"), or some mix of these.[209]

In any case, my earlier remarks about using the *Agent_Zero* framework to study financial contagions, the collapse of spatially explicit housing markets, or other cascading economic dynamics is far from idle. Given relevant data, the replication of historical episodes—economic and otherwise—could profitably be attempted. This may inspire the collection of relevant data, both economic and neurocognitive.

III.14. SPIRALS OF MUTUAL ESCALATION

As a concluding extension, I would like to "close the circle," harking back to our opening examples of violent occupation by Blue *Agent_Zero* actors. Thus far, the patches—interpreted as a yellow indigenous population—have not been endowed with any *agency*. They simply go active (turning orange) at an exogenous user-specified rate. And, thinking of the opening civil violence examples, they don't *retaliate* on their occupiers in any way. They can't "shoot back," as it were. Here, I will add both these elements of agency—adding an endogenous attack (i.e., rebellion) rate and the ability to return fire—in the simplest way. This will permit us to generate mutual escalation spirals.

Of course, both colonial and military history are full of cases in which repressive force by occupying powers has exacerbated underlying grievances. The occupier destroys the landscape and infrastructure and kills innocent civilians. This backfires and produces more rebels. These new insurgents inflame the hostile passions of the occupiers, who respond with more indiscriminate violence, producing more civilian casualties, which begets further rebels, in an upward spiral of mutual escalation. One has only to think of the current (2012) events in Syria as an immediate example, though history is replete with others. Sometimes the rebels prevail, and sometimes they are wiped out. These spiral dynamics and various outcomes can be quite naturally generated from the core model with a few simple alterations.

Endogenous Attack Rate and Retaliation

As before, the Blue occupier's affect grows (à la Rescorla-Wagner) with aversive stimuli (rebel attacks). We apply the earlier-developed extension

[209] One might explore the associative strength between consumption (the CS) and pleasure (the CR) as a basis for so-called brand loyalty. Similarly, so-called umbrella branding may exploit stimulus overgeneralization, so that merely bearing the name Porsche or Sony confers credibility, even if the product is sunglasses, not cars or TVs. I thank Robert Axelrod for this suggestion.

wherein the occupier's damage radius increases with his affect.[210] As this damage radius increases, so does the level of general destruction (including *innocent* civilians). Now we introduce our two extensions.

First, the *attack rate itself* (the frequency with which quiescent yellow patches go active, turning orange), which had been constant,[211] now grows with the level of destruction.[212] But this endogenous rebel attack rate produces a feedback effect: the higher indigenous attack rate produces a *higher* occupier affect, in turn producing a greater damage radius and more indiscriminate destruction. This produces a higher insurgent attack rate, inspiring more indiscriminant destruction, more insurgents, and so on, in the upward spiral. But who wins?

Thus far, the yellow patches have not been permitted to return fire. So, the second extension is to now assume that when a patch goes rebellious (turns orange), it has some *user-specified probability*[213] *of killing any occupier within range.*[214] We arm the patches and let them return fire. Mao's famous dictum was that revolutionaries must "swim like fish in the sea," vanishing

[210]To some extent things will escalate without this. Since the attack rate grows endogenously, the occupiers are getting more aversive stimulus, so their disposition exceeds threshold more frequently. So the frequency of attacks, but *not* their destructive radii, will grow in any case.

[211]Thus far, it has been a constant set by the user, using the attack_rate slider in the model's *NetLogo* interface.

[212]Technically, before, we used total disposition. Here, it is pure affect that grows endogenously. The level of destruction is the number of dark red patches (colored red-3). The attack rate begins at a low background level of 10 and then updates endogenously as follows:

```
to update_attack_rate3
  ifelse count patches with [pcolor = red – 3] = 0;
    [set attack_rate3 10]
    [set attack_rate3 10 + 100 * count patches with [pcolor = red – 3] / 1089]
end
```

The constant 1089 is the default number of patches in the *NetLogo* interface. Division by 1089 thus places an upper bound of 110 on the attack rate in this extension. For complete code, see the Princeton University Press Website.

[213]We assume these are uniformly distributed for expository purposes.

[214]In the base run, I assume this range to be the von Neumann neighbors. Think of this as the patches' vision. The per-shot probability of killing an occupier is 0.05. The specific code block is:

```
to retaliate-orange
  ask patches
    [
    if pcolor = orange + 1 and any? turtles in-radius 1 and random 100 < 5
    [ask turtles in-radius 1 [ die ]]
    ]
end
```

As always, full code is included in the Applet on the Princeton University Press Website.

into the general population (returning to the yellow appearance) after each openly rebellious act. Our rebels will do likewise.

These alterations suffice to generate upward spirals and the core result in which, unlike the opening parable of unbridled rampage, the rebels succeed in defeating the occupiers. As an existence proof (not a full robustness analysis), I offer the run of Figure 112.

Panel 1 (upper left) shows sporadic opposition activity before any destruction by occupiers. As destruction progresses, the rebellious attack rate itself grows, as shown in panel 2 (upper right). However, the Blue

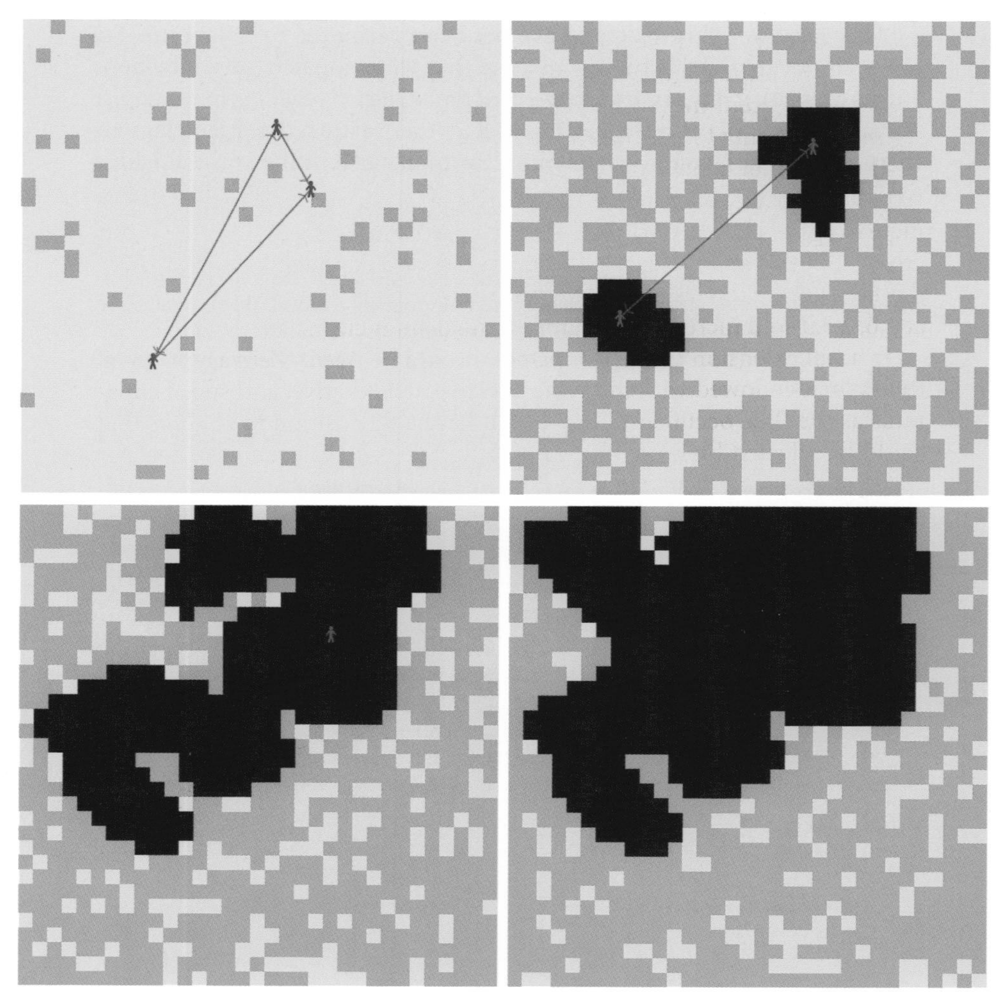

FIGURE 112. Successful Revolution [Movie 14]

destructive radius also grows as the frequency of insurgent attacks increases. When their destructive radius reaches 2, the occupiers can destroy sites without actually stepping on them—from inside a protective radius—as also shown in panel 2. Notice that, operating from within these enclaves, the Blue agents are not subject to any direct stimulation. So, their affect does not update in the Rescorla-Wagner model. They continue killing with *literally flat affect*. However, the insurgents are assumed to have weapons that can reach 1 unit into these enclaves. So, when a Blue gets near an edge, he is vulnerable to the insurgent orange sniper (who, as per Mao, turns yellow again with some probability). One of the Blue occupiers has been picked off in panel 2, and a second in panel 3 (lower left). Finally, with half the country destroyed, the occupiers are thwarted (panel 4). At this point, most of the country is radicalized (orange). The remaining yellows are a mix of recent members of *la resistance* and remnants of the *ancienne regime* pretending to be. The entire revolution is shown in **Movie 14**.

The insurgent attack rate has grown throughout, as evidenced by the increased frequency of orange sites. The plot of attack rate over time is clearly increasing, as shown in Figure 113. It rises from 10 to 52, a 5.2-fold increase. But this and the damage radius are closely coupled. In fact, tracking the Blue agent who lasts longest, his damage radius begins at 0.3 and ends at 1.6, a 5.3-fold increase. Escalatory mutualism is clear.

In a full extension the yellow patches would be *Agent_Zero* agents in all their glory, endowed with the same affective, deliberative, and social apparatus as the Blue occupiers. They would have a disposition *to retaliate* that could be affected by homophily dynamics (shared antioccupier affect for example) and endogenous network structure, as explored earlier. This full generalization is suitable for future research. Here, however, is a simple step

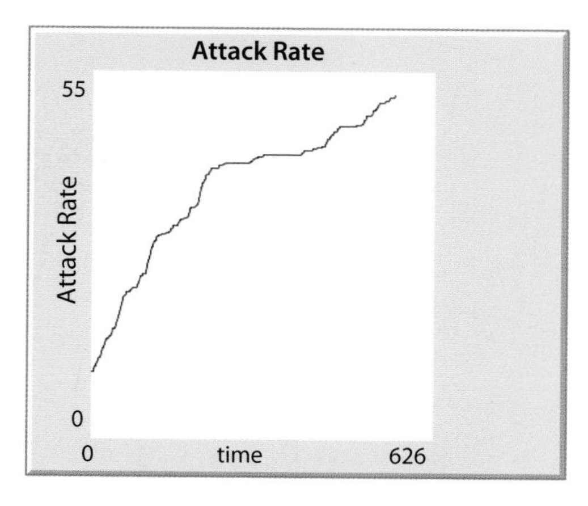

FIGURE 113. Endogenous Rebel Attack Rate

in that direction which—by endowing the patches with minimal agency—generates the upward mutual escalation dynamic observed in many rebellions and anticolonial struggles. In addition to endowing all patches with agency (making them all *Agent_Zero*s), numerous lines of future research strike me as fertile.

PART IV

Future Research and Conclusions

IV.1. FUTURE RESEARCH

In this exposition, I have been essentially interested in the feasibility of a fundamental synthesis, rather than in its robustness to numerical perturbations. Hence, I have left sensitivity analysis for future rounds of work. But it would certainly be interesting to more fully explore the parameter space of the model, charting out its regions of stability.

Numerical Cartography

This numerical cartography could naturally begin with the systematic covariation of the (currently) global variables[215] offered as sliders in the *Agent_Zero* interface. These are as follows:

- Attack rate (stochastic environmental stimulus rate)
- Spatial sampling radius (sometimes referred to advisedly as "vision")
- Extinction rate
- Memory length
- Action radius

The existing code also allows users to make a number of binary choices:

- Moving average or moving median
- Probability judgments biased by affect, or not.
- Weights exogenous or changing endogenously through strength-scaled homophily
- Homophily based on affect or on probability
- Fight vs. flight.

[215] Global variables are the same for all agents.

Greater Heterogeneity

Agents are already heterogeneous by action thresholds, weights, and learning parameters, for example. But variables and binary settings that are currently global (identical across agents) can also be made to differ across agents. Some agents could use a moving average over a short memory window, while others employ a moving median over a long memory; for some, probability is biased by affect, for others, it is not. Some agents might have high spatial sampling radii, while others are effectively blind.

Tool Set

When one considers the enormous number of possible combinations constructible by covarying these assumptions, it is clear that this book merely scratches the surface of what is constructible from the existing *Agent_Zero* code. Innumerable explorations and sensitivity analyses can be pursued by adapting the code provided in Appendix III and the full collection of *Mathematica* and *NetLogo* programs posted on the Princeton University Press Website. This tool set allows for much further research.

Modular Variants

As noted earlier, the underlying affective, deliberative, and social network components—the submodels of *Agent_Zero*—are all open for discussion and refinement. This entire development could be seen as a single instance of the myriad models one could assemble from different affective, deliberative, and social constituents, coupled in various ways, as suggested in Table 2. The base *Agent_Zero* model is the top row.

For instance, one could use the Pearce-Hall model (Pearce and Hall, 1980) rather than Rescorla-Wagner for the affective module. We have ignored anchoring affects in the estimation of probabilities; they could be

Table 2. Modules for *Agent_Zero* Variants

Affective/Emotional Module	Rational/Deliberative Module	Social Structure (Directed Weights)	Executive Function
Rescorla-Wagner	Local sampling and relative frequency	Fully connected	Summation
Pearce-Hall	Anchoring	Dynamic connectivity	Sigmoid of summation
Neural network	Bayesian updating	Small world	Multiplicative rather than additive

included in the deliberative module. We explored both fully connected and dynamic network structures. But one could explore many others, including small-world networks (Watts and Strogatz, 1998). *Agent_Zero*'s overall disposition is the sum of emotional, cognitive, and social components. Future elaborations of the model could explore alternatives to this summation, as in some of the neural network literature.

Indeed, while affect, probability, and social influence are all dynamic, *at any particular time* the unadorned *Agent_Zero* of Parts I and II is, algebraically, a *perceptron* (Rosenblatt, 1962). That is, the dot product of the input and weight vectors is compared to a threshold.[216] One obvious move from the feed-forward neural network literature is to process this raw dot product—the weighted sum of components—running it through a sigmoid function, as in a back-propagation network [see Rumelhardt and McClelland (1987) and Freeman and Skapura (1991)], as suggested in column 4.

One could think of *Agent_Zero* as a template for a modular social science, where one can swap in and out various affective, deliberative, and social modules.[217] And, again, agents could be heterogeneous in these compositions.

Proximity-Dependent Weights

Earlier, I noted that the spatial sample radius employed in the estimation of probability (P) was local, and mathematically independent of the (unbounded) range at which dispositional contagion can, by myriad means, occur. This independence means that Agent 1's weight on Agent 2 is no different when 1 is inside 2's sample radius (his "vision," as it were), than when she is not.

Now, one might wish to explore the consequences if the weight of Agent 1 on Agent 2 is greater when Agent 1's binary behavior *is* observed by Agent 1 (occurs within the sampling radius) than when it is not. Since action, A, is either 0 or 1, an obvious way to introduce this would be to generalize the weights as follows: for agents j within i's sample radius, the new weight ω^*_{ji} is given by

$$\omega^*_{ji} = \omega_{ji} k^A, \qquad [61]$$

[216]This is the input to the Heaviside function first introduced in Part I.

[217]As noted earlier, the fact that the *Agent_Zero formalism* can be modular in this technical sense has nothing to do with whether the human brain is "modular," however one may define the term. The model need not "look like" the thing being modeled, in other words. While a scale model of a bridge does resemble a bridge, I'm not sure it even makes sense to say that a mathematical model "looks like" the thing, or process, being modeled. Hooke's law "looks" nothing like a spring. It looks like $F = -kx$. The spring oscillates, the law doesn't. The equations of general relativity govern, but do not resemble, a galaxy.

where real $k \in [0, 1]$ alters the original weight ω_{ji} when $A = 1$ and does not when $A = 0$. In the pure dispositional contagion model, $k = 1$ (which of course stays 1 for either A). We might assign someone high weight based on a cell-phone exchange with her, but when I meet her in person, I see that she is of an unexpected race (or age) and her weight drops. Or a person might have a certain weight based on text exchanges, but when I hear his sophisticated foreign accent, his credibility immediately goes up. In effect, it is a stereotyping extension.[218]

Neural Deepening

As neuroscience progresses, it may be feasible and productive to increase the modeling resolution, making *distributed brain complexes themselves the agents*, and our *Agent_Zero* constituents—affective, deliberative, and social—the emergent phenomena, as it were.[219] Agent-based models have generally taken the individual person as the agent. We have introduced affective, cognitive, and social modules. But, we may be able to increase the magnification further, make internal neural complexes themselves the agents, and grow these components. For example, with regard to outright addictions—to tobacco, alcohol, drugs, and foods, for example—the affective module could be deepened to represent the action of relevant neurochemical reward systems. This project appears to be well underway for the dopamine reward system and its role in addiction (e.g., Glimcher, 2011).

Scale-Up

While deepening the agents as the neuroscience warrants, I plan to scale the space and the agent population up dramatically. The feasibility of planetary-scale agent-based modeling is demonstrated in Parker and Epstein (2011). There, the Global Scale Agent Model (GSAM) simulates infectious disease dynamics on a planetary scale, efficiently running 6.5 billion distinct individuals (Parker and Epstein, 2011; Epstein, 2009). But it is easily adapted to the analysis of other large-scale social dynamics. The combination of higher neural resolution and vastly larger scale is an important scientific project.

However, the scale-up, while technically feasible, will need to capture an important phenomenon—namely, that the impact of others' experience on us often exhibits diminishing marginal returns. In the three-agent version developed here, each agent's total disposition equals simply her own solo disposition ($V + P$) plus the sum of others' weighted solo dispositions. We see no diminishing marginal impact as the *number of influencers* grows.

[218] I thank Julia Chelen for this interpretation.

[219] For a spirited critique of classical emergentism, see J. M. Epstein (2006, Ch. 2).

For two influencers this may be defensible. But when the numbers get larger, unadulterated addition probably breaks down. In studies of Latané (1981), for example, impact (assuming unit weights) was found to scale log-linearly in numbers, with slope less than 1. Slovic's work on psychic numbing (2007) gives even more dramatic scaling examples from such phenomena as famine and genocide. Millions the world over can be riveted by the drama of 33 miners trapped underground (as in Chile 2010), while millions trapped in famine or genocide are treated with virtual indifference. Something tragic happens when we scale up. Empathy is dramatically subadditive. And a large-*n* model will need to represent that, and perhaps point to remedies for it.

Empiricism

Nicholas Kaldor introduced the term *stylized fact* into economics, and it is taken as evidence of empirical credibility when economic models are shown to generate stylized facts of economic growth or business cycles, for example. I would make the same sort of claim for *Agent_Zero*, except that in our case, they have been "stylized dynamics," of mutual escalation, for example. A somewhat stronger empirical claim might be made for our replication of Latané and Darley's experiment.

Where data permit, it would be important to attempt the kind of full agent-based computational reconstructions that have been conducted for the Anasazi, for epidemics, and other phenomena (J. M. Epstein, 2006). Where data do *not* yet permit, theoretical models like *Agent_Zero* can guide its collection. This is an underappreciated role for models. Frequently, theory precedes and guides data collection (J. M. Epstein, 2008), and I welcome further empirical work on dispositional contagion.

Specific Hypotheses

Models also contribute to empirical activity by furnishing testable hypotheses. Of course, the broadest hypothesis is that *Agent_Zero* provides a generative explanation of many important phenomena. But, more specific hypotheses have also been tentatively advanced along the way. Among them are these:

1. That the mechanism of *threshold imputation* can explain the Latané-Darley and related results in social psychology;
2. That *affect* can amplify probability judgments log-linearly;
3. That the probability of network connection between two agents varies with the product of aggregate affective strength (the sum of affects) and homophily (one minus the absolute affective difference),

which offers an alternative to preferential attachment as a mechanism of network formation.

I would welcome experimental or empirical work along all the preceding lines. Meanwhile, I hope that the versatility and extensibility of the *Agent_Zero* model have been shown. It is an initial step in the direction of a more unified and neurocognitively grounded computational social science. While looking forward to empirical exercises, my claim would be that the simple present model is *sufficient to generate* important dynamics, that it "gets at" deep things and rings true, perhaps in the way literature does.

Literature

Indeed, as noted in the Foreword, my previous book, *Generative Social Science*, ends with a challenge: *Grow Raskolnikov*. Most of our serious dialogues—like Raskolnikov's own—are with ourselves. And inwardly, I think this is actually what I've been working on since throwing down that gauntlet in 2006. *Agent_Zero*, of course, is not Raskolnikov, but he is recognizable in the same way, and is a fruitful ideal type. Obviously, the scientific merit of *Agent_Zero* does not depend on the analogy to Dostoevsky's character. And, as noted, the invocation to "Grow Raskolnikov" was really about growing an internally conflicted agent. Despite that choice being somewhat offhanded, I have enjoyed exploring this parallel and see no harm in discussing it's appropriateness after all. Like *Agent_Zero*, Raskolnikov is possessed of (indeed possessed *by*) distinct and competing modules: (a) an abstract intellectualized one, in which he conceives the murder of his decrepit, but innocent, pawnbroker, (b) an emotional one, in which he experiences the most profound self-loathing over this impulse, and (c) the diffuse influence of the nihilist social movement spreading across educated Russian society (Raskolnikov is a former student) at the time. In his case, the nihilist and intellectual modules prevail over his revulsion, and he commits murder. Recall that in Part III, we conducted an *Agent_Zero* experiment in which the destructive radius is endogenous, and—with sufficiently poor impulse control—can exceed the stimulus squares. Raskolnikov's murder radius exceeds the stimulus also: he impulsively kills the pawnbroker's innocent sister Lizavetta, who happens to appear at the crime scene.[220]

Obviously, the scientific success of the model does not rest on the success of the literary analogy. But I am gratified that my choice of Raskolnikov was felicitous. It would be very interesting to grow other great figures in literature, or social plot lines, enduring parables that they are!

[220] One could argue that the second murder is "rational," in eliminating a witness. But Raskolnikov does not make this calculation; he kills impulsively.

Data, after all, is the poor man's literature. But, in all sobriety, an important line of future research will be to compare instantiations of the model—for violence, contagious fear, flight, economic dynamics, jury deliberations, or health behaviors—to available data.

IV.2. CONCLUSIONS

This volume introduces a new theoretical entity, *Agent_Zero*, whose disposition and behavior depend on affective, deliberative, and social modules. Relevant neuroscience was discussed throughout. Mathematical and agent-based versions were developed. The agent version is explicitly spatial—space is a landscape of stimuli. In many interpretations, yellow patches are "good," while orange ones are "bad," or aversive. Agents process the orange stimuli as if they were trials in a conditioning dynamic governed by the (generalized) Rescorla-Wagner model. This produces their affective trajectory $V(t)$. At the same time, they compute the local relative frequency of aversive stimuli and interpret this as the probability that a random patch is an aversive (e.g., enemy) patch. They have memory and in the present version can employ a moving average or moving median over their sampling window. This produces their deliberative trajectory $P(t)$. The emotions and deliberations of others influence each individual's total disposition, $D(t)$, which is the sum of weighted solo dispositions, taken over the individual's bi-directional network, where self-weights are unity. This produces what I have dubbed *dispositional contagion*, as distinct from behavioral imitation.

Interagent weights are endogenized as functions of affective homophily in the model. Each agent has an action threshold, τ. If her total disposition exceeds it, the agent takes the binary action A in question. Succinctly, $A = H(D - \tau)$. It is conceptually a very simple model with three parts: affective, deliberative, and social. But it was shown to have considerable generative power. In the course of this exposition, a wide range of interpretations has been developed. I review some here. The first context is the darkest.

Civil Violence

Slaughter of Innocents

In one computational parable, *Agent_Zero joins* a lynch mob (or genocide) despite having no aversion to black people, no evidence of black wrongdoing and no orders to engage in violence. In another, *Agent_Zero initiates* the lynch mob, not because he is in fact a "leader" in the usual sense, but only because he is the most susceptible to dispositional contagion. If we interpret *Agent_Zero* as a soldier occupying a foreign country, the stimuli are enemy ambushes and the damage radius is the area (the yellow population) against which *Agent_Zero* indiscriminately retaliates. This is the My Lai massacre, or the Nazi "collective reprisals," in which whole villages were annihilated in retaliation for isolated resistance actions.

Overall Picture

The overall picture of Homo sapiens reflected in these interpretations of *Agent_Zero* is unsettling: Here we have a creature evolved (that is, selected) for high susceptibility to unconscious fear conditioning. Fear (conscious or otherwise) can be acquired rapidly through direct exposure or through observation of fearful others. This primal emotion is moderated by a more recently evolved deliberative module, which, at best, operates suboptimally on incomplete data, and whose risk appraisals are normally biased further by affect itself. Both affective and cognitive modules, moreover, are powerfully influenced by the dispositions of similar—equally limited and unconsciously driven—agents. Is it any wonder that collectivities of interacting agents of this type—the *Agent_Zero* type—can exhibit mass violence, dysfunctional health behaviors, and financial panic?

Computational Doppelganger

Of course, the central human paradox is precisely that we do see another side: the same Germany that produces Hitler produces Einstein. And, for every ten stories of collective brutality, there are some (fewer than ten, I suspect) stories of collective resistance to it.

Arab Spring

Interpreting orange stimuli as instances of Arab regime corruption and agent actions as the rebellious removal of illegitimate authorities, we generated a caricature of the 2011 Arab Spring. The crucial role of social media in enabling affective homophily to amplify ties and dispositions was demonstrated. Indeed, we constructed a pair of runs in which the same agents, without social media, do not rebel. But, propelled by the dispositional amplification afforded by it, they do. These same stylized facts, of course, characterize a wide range of insurrections, whose empirical reconstruction would be important.

Economics

Contagion and Capital flight.

In the preceding examples, the agents' action is essentially destructive. They wipe out a village, depose a regime, or intern an ethnic group. They fight.

But the model easily generates the flight response. One such interpretation posits a landscape of financial or property assets. The agent's portfolio is the set of assets within his "vision." Yellow assets are healthy (making acceptable returns). If they suddenly lose value, they turn orange. This results in changes in the agent's fear (*V*) that asset values will tumble and changes in his computed probability (*P*) of such events. If these conspire to elevate disposition above threshold, the agent flees his portfolio, a disposition that can cascade through network effects.

Price Dynamics and Seasonal Cycles

Prices were explicitly introduced into the threshold term. The disposition (or propensity) to consume depended on price in an orthodox fashion. Cycles of supply (as in seasonal variations in strawberry production and transportation costs) were easily generated. Cycles of demand (as in Christmas toy sales) were generated by allowing the landscape of stimuli to oscillate, which drove dispositional cycles. So we did more than simply legislate that demand would oscillate. Affective and dispositional dynamics were intervening forces.

Marketing

Various marketing strategies were shown to have analogues in the mobilization of affective, deliberative, or social components of the agent model. Some further extensions (e.g., *Agent_Zero* as CEO) were sketched.[221]

Health Behavior

Vaccine Refusal

If the space is a landscape of vaccines, the behavior of interest becomes vaccine refusal, an area rife with emotionality. Given an orange event—an adverse event, or even a report of one—with one vaccine, some agents foreswear all vaccines within a large pharmaceutical radius, while others refuse only a narrow set. As in earlier research (J. M. Epstein et al., 2008), fear itself can, of course, be contagious and generate a large swath of vaccine refusal, far beyond what an empirical risk deliberation would warrant.

Unhealthy Eating

Binge eating is another health interpretation where the space could be coordinatized by carbohydrates (x-axis) and fat (y-axis). Despite computing

[221] Brand loyalty was cast as a conditioning phenomenon, and it was suggested that "umbrella branding" exploits stimulus overgeneralization.

a high probability of unhealthy effects, the agent—perhaps driven by social network effects—can impulsively consume a large radius of foods, once he is conditioned to associate them with pleasure, holidays, or membership in a group.[222] Indeed, as we reviewed, the neuroscience (Kross et al., 2011) suggests that resistance to the group norm can be literally painful.

Psychology

Aging and Impulse Control

We showed how the destructive radius could be endogenized. Then we made it a function of affect and age, to model the well-known fact that, typically, impulse control is lower in minors than in adults, as recognized in legal distinctions between the two, for example.

Latané-Darley and Zillmann

We replicated the famous Latané-Darley smoke-filled room experiment from social psychology, proposing a new explanatory mechanism—threshold imputation. We also showed the model to be consistent with Zillmann's experiment, in which connection we ventured another testable hypothesis that affect biases probability in the form $P_e = P_n^{1-V}$, which is to say log-linearly in $1 - V$.

Posttraumatic Affect Retention

Agent_Zero's affect can persist undiminished long after any true stimulus has stopped. A variety of simple recognizable parables were generated: bringing a bad mood home from the office, or bringing traumatic fear home from combat or a fire rescue, despite the stimulus episodes ending long before. Passion and reason can move on different time scales. In connection with posttraumatic stress persistence, we explored the effect on the group if even one agent is unable to reset her λ to zero.

[222] In a more general energy-balance interpretation, the x-axis could be caloric intake; the y-axis could be level of physical activity, and the z-axis could be BMI. Agents located in the lower right are sedentary (low y) and consume high calories (high x). These have high BMI. One wishes to move them to a low-x/high-y lifestyle. But certain paths are clearly not feasible. Morbidly obese sedentary agents cannot immediately increase y vertically and then go left to a low x. They may be physically incapable of immediate vigorous exercise, or their current (e.g., inner-city) environment may preclude it. So, if orange outbursts are opportunities for unhealthy behavior, one wants to present them with a "yellow brick road" to a healthier location. But, with heterogeneous affective and deliberative modules, different social networks, physical opportunities, and budgets, successful trajectories may vary widely across agents. For a variety of simple diet trajectories that could be tailored to heterogeneous individuals, see Hammond and Epstein (2007).

Knockout Agent

We conducted what I believe to be the first in silico lesion study, showing that the excision of one agent's emotional module had effects not only on their affective trajectory and behavior, but also on their capacity to transmit affect to others and hence had systemic ramifications at the group level.

Jury Dynamics

Three-Phased Process

We used the framework to model a three-phased trial process. Phase 1 was pretrial, in which agents are bombarded with claims of guilt (orange) and presumptions of innocence (yellow) in the public square. They form a V and a P about the defendant before any formal trial begins. Then, if they are chosen as jurors, they are subject to an entirely different stimulus pattern in the courtroom. Their V and P evolve accordingly. Thus far they have still not interacted with other jurors. When, at last, they are sent behind closed doors to deliberate collectively, they reveal their dispositions to convict; network and momentum effects well documented in the literature are then generated. Overall, the verdict under jury deliberation can be entirely different from what any individual juror in isolation would deliver. Counterintuitive effects of a change of venue were discussed.

The Formation and Dynamics of Networks

Affective Homophily and Endogenous Weights

Suffusing all of this are the interagent weights. Initially these weights are exogenous. Then they are endogenized as the product of affective strength $(v_i + v_j)$ and affective homophily $(1 - |v_i - v_j|)$. Literature supporting this idea is cited. Then the binary formation of links is modeled as the Heaviside step function of connection weight minus a link threshold. This is an alternative hypothesis to preferential attachment and is yet another testable hypothesis given relevant data. The model can easily accommodate affective, probability, or disposition homophily rather than affective, which was solely explored here.

Multilevel Societies

We outlined how the model could be naturally extended to generate hierarchies in which the *actions* of Layer_n agents are treated as *stimuli* by Layer_$(n + 1)$ agents. A number of interpretations—regulatory and peacekeeping organizations—were discussed.

Mutual Escalation Dynamics

In all the preceding discussion, the yellow activation rate was an exogenous constant set by the user. The concluding extension endows the yellow patches with agency. Harking back to the opening examples of civil violence, we give yellow patches the abilities to (a) endogenously increase their attack frequency in response to occupier destruction and (b) to retaliate on occupiers. We also incorporated a variant on the first of the previously noted extensions, making the occupiers' destructive radius an increasing function of affect. These alterations generated escalation spirals and permitted the resistance to defeat the occupiers, a general dynamic observed throughout political-military history.

Birth and Intergenerational Transmission

Taking as our inspiration a famous historical passage from Marx's *The 18th Brumaire of Louis Bonaparte*, we developed an intergenerational parable in which the parent's memory—her chronicle—is initially imprinted on the child. But the child leaves home, has new experiences, and eventually "overwrites" the inherited narrative with his own.

We generated all this without any assumption of *behavioral* imitation. It would appear that a fairly rich social science might be possible without it!

IV.3. TOWARD NEW GENERATIVE FOUNDATIONS

Scientific theory should *not* aim at realism. It should aim at fruitful idealizations, from which real entities and phenomena can be productively cast as perturbations. There are no ideal gases or frictionless planes. But these limiting cases—which do not occur in nature—turn out to be the productive theoretical entities.[223] Not all limiting cases do.[224] *Agent_Zero* is such an idealization.

Agent_Zero vs. *Homo Economicus*

His or her construction has been my central objective. Recall our opening example. With no negative affect and no relevant evidence, he or she yet perpetrates—indeed initiates—acts of destructive violence. Do real people

[223] For an incomparable statement of this Cartesian perspective, see Joseph Epstein's Introduction to *A Discourse on Method, and Other Works* (Descartes, 1965; J. Epstein, ed.).

[224] The physically accurate limiting case for our solar system is heat death. But this would be an unproductive starting point for the study of biology on Earth.

behave exactly that way? Hopefully, not many. But, if one wished to model participants in genocide, which would be the better limiting case—*Agent_Zero* or *Homo economicus*? Which ideal type most naturally accommodates the recent insights of cognitive neuroscience (e.g., indirect fear contagion) or the robustly documented logical confusions, elemental conformity effects, and contagious hysterias observed in *Homo sapiens*? Certainly, in these contexts, *Agent_Zero* has some claim to primacy.

It is a simple unified neurocognitively grounded model able to crudely generate central phenomena across the spheres of conflict, economics, health behavior, law, social psychology, and network dynamics. It has what I might call high *generative efficiency*. The *Agent_Zero* model itself is minimal, but it generates an extensive space of phenomena. We've squeezed a lot out of it, in other words.

Whether this particular agent, or some distant progeny yet to emerge, proves the most enduring, I believe this broad family tree of individuals—each capable of emotional learning, bounded rationality, and social connection—is well worth developing. With agent-based modeling, large numbers of heterogeneous agents in this family can interact directly with one another, generating interdependent dispositional and behavioral trajectories in time and space, which can be compared to data and used to better understand—and perhaps improve—a wide range of important social dynamics. In sum, while it can surely be refined, I offer *Agent_Zero* as a step toward neurocognitive foundations for generative social science.

Appendix I.
Threshold Imputation Bounds

LET US GENERALIZE THE SKELETAL EQUATION for Agent 0 by introducing a constant ψ as follows:

$$D_0^{net} = V_0 + P_0 + \omega_{10}(V_1 + P_1 - \psi) - \tau_0. \quad [1]$$

We interpret ψ as the threshold Agent 0 imputes to Agent 1. In Parts I and II, in the canonical run where Agent 0 precedes Agent 1, no such variable appears: implicitly, $\psi = 0$ and no imputation is assumed. As an extension, in Part III, we showed the Darley experiment to be generable if, in that setting—where Agent 0 *can observe* the action of others—Agent 0 is assumed to impute a positive threshold $\psi > 0$ to Agent 1.

Jon Parker made the acute observation that if the imputed ψ happens to equal exactly the true τ_1, we cannot simultaneously satisfy the conditions of both the canonical run and the Darley run.

Specifically, recall that in the canonical run where Agent 0 experiences no direct stimuli (attacks), $V_0 = P_0 = 0$. If Agent 0 is to act, we need

$$\begin{aligned} &\omega_{10}(V_1 + P_1 - \psi) > \tau_0 \text{ implying that} \\ &(V_1 + P_1 - \psi) > 0. \end{aligned} \quad [2]$$

But, if ψ equals exactly the true τ_1 and Agent 1 is *not* acting, it must be the case that

$$(V_1 + P_1 - \psi) < 0. \quad [3]$$

Clearly, equations [2] and [3] cannot not both obtain. So, for what range of ψ values are both the canonical and Darley runs generable?

Claim: The condition is: $\tau_1 > \psi > 0$.

Proof: With $V_0 = P_0 = 0$, Agent 0 acts if $\omega_{10}(V_1 + P_1 - \psi) > 0$, or simply

$$V_1 + P_1 - \psi > 0. \quad [4]$$

Agent 1 does *not* act if $V_1 + P_1 - \tau_1 < 0$, or multiplying both sides by -1,

$$-V_1 - P_1 + \tau_1 > 0. \quad [5]$$

Adding [4] and [5] we obtain $\tau_1 - \psi > 0$. So, with

$$\tau_1 > \psi > 0 \qquad [6]$$

we can generate both the canonical and Darley results. QED.

Hence, the core run—in which Agent 1 goes first—is *more* robust than suggested, since it holds up under a range of ψ values, not just zero, as was assumed in the earlier exposition.

This argument has been for two agents but obviously will generalize to arbitrary n.

Appendix II.
Mathematica Code

II.1. Complete Solution of the Rescorla-Wagner Model in Step Functions

The following *Mathematica* 8.0 code implements equation [17], the complete acquisition and extinction trajectory for the Rescorla-Wagner model, using Heaviside unit step functions. The subsequent Animate code generates **Animation 0** of this entire trajectory as the parameter, n, is varied from 0 to 30 in increments of 0.1. A snapshot ($n = 10$) from the movie is shown first. The full movie is posted as **Animation 0** on the book's Princeton University Press Website.

```
Plot[UnitStep[10 - t] (1 - Exp[-.15 t]) + UnitStep[t - 10] (1 - Exp[-1.5])
    Exp[-.15(t - 10)], {t, 0, 65}, PlotRange → {0,1}]
```

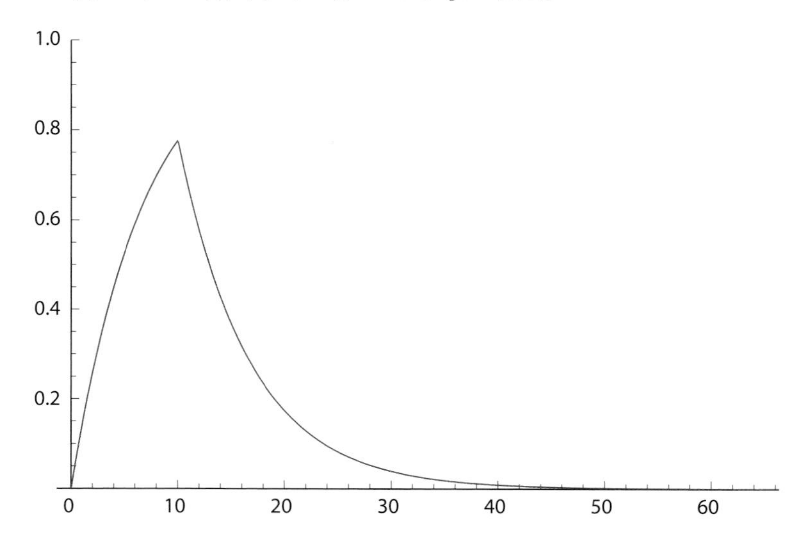

```
Animate[Plot[UnitStep[n - t] (1 - Exp[-.15 t]) + UnitStep[t - n]
          (1 - Exp[-.15 n]) Exp[-.15 (t - n)], {t, 0, 65}, PlotRange →
          {0, 1}], {n, 0, 30, .1}]
```

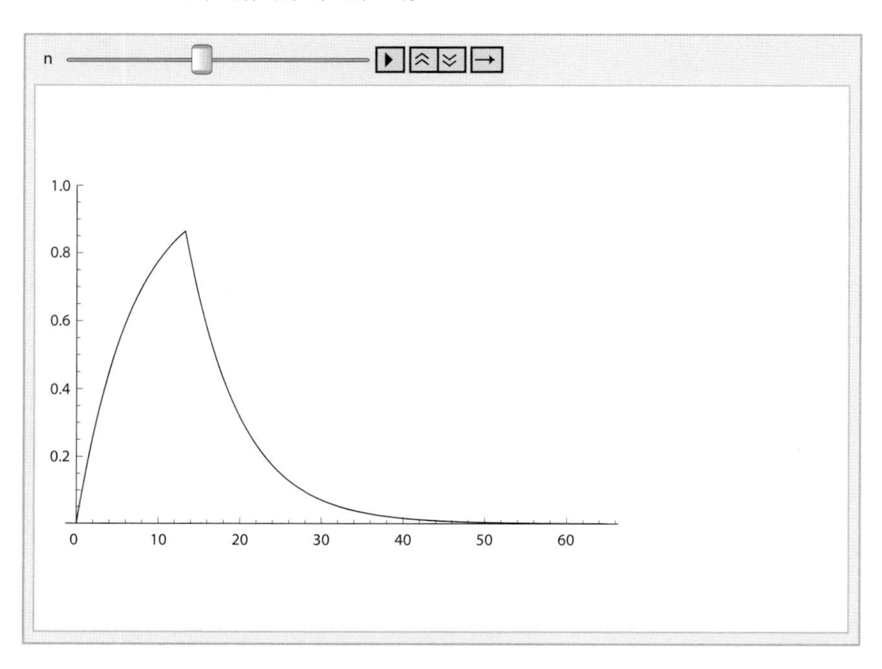

II.2. Two-Agent Dispositional Contagion: From Negative Disposition to Inititation of Action

The Four Panels of Figure 18 (includes Figure 17)

This entire section (and the next) can be compressed into a single *Mathematica* Module. But for nonprogrammers, a step-by-step exposition will be clearer. We begin by defining three global constants (probabilities and a common threshold) that will be used throughout. Note that Agent 1's probability is zero. Then we use Mathematica's nonlinear ordinary differential equation solver, NDSolve, to generate a numerical solution to the Rescorla-Wagner equations, with initial affect equal to zero for both agents. The S-curve exponent, delta, is also zero. *Mathematica* reports when it has computed Interpolating Functions, a message we shall suppress. Then we vary the weight of Agent 2 on Agent 1 from 0 to 0.9, showing the plots in each case. The pair of plots given in Figure 17 are from the 0.7 case shown next.

```
p1 = .00
p2 = .80
Tau = 1.0

rw15 = NDSolve[{
            v1'[t] == 0.06 (v1[t]^0) (1 - v1[t]),
            v2'[t] == 0.02 (v2[t]^0) (1 - v2[t]),
            v1[0] == 0, v2[0] == 0},
            {v1, v2}, {t, 650}]
```

0 : Never positive

```
Plot[Evaluate[{v1[t] + p1 + .0 (v2[t] + p2) - Tau,
              v2[t] + p2 + .4 (v1[t] + p1) - Tau
              } /. rw15], {t, 0, 150}]
```

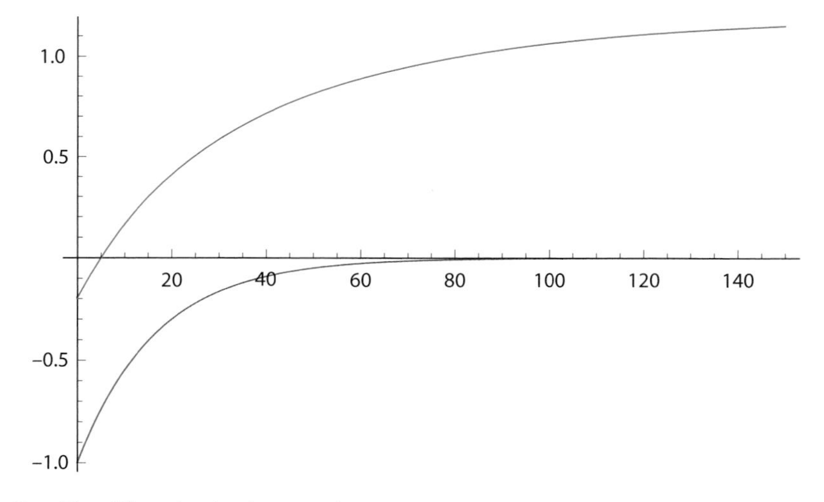

.3 : Positive but always lower

```
Plot[Evaluate[{v1[t] + p1 + .3 (v2[t] + p2) - Tau,
              v2[t] + p2 + .4 (v1[t] + p1) - Tau
              } /. rw15], {t, 0, 150}]
```

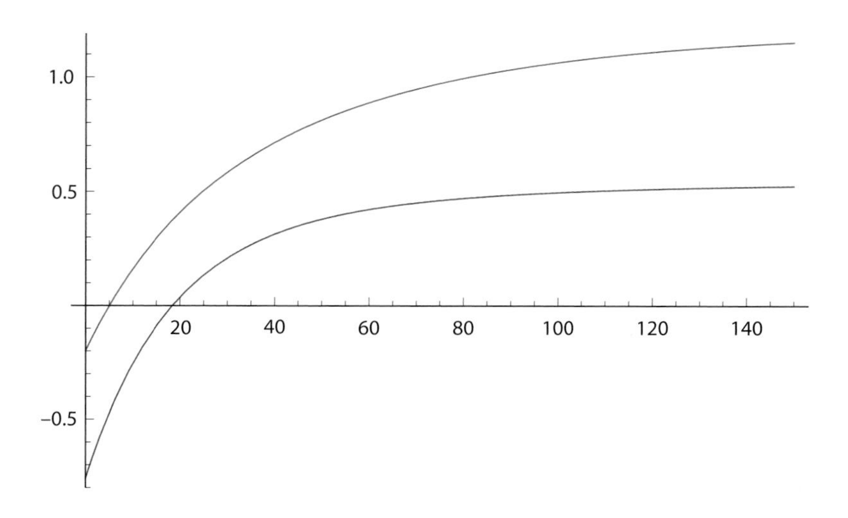

.7 : Supasses but goes second

```
Plot[Evaluate[{v1[t] + p1 + .7 (v2[t] + p2) - Tau,
               v2[t] + p2 + .4 (v1[t] + p1) - Tau
               } /. rw15], {t, 0, 50}]
```

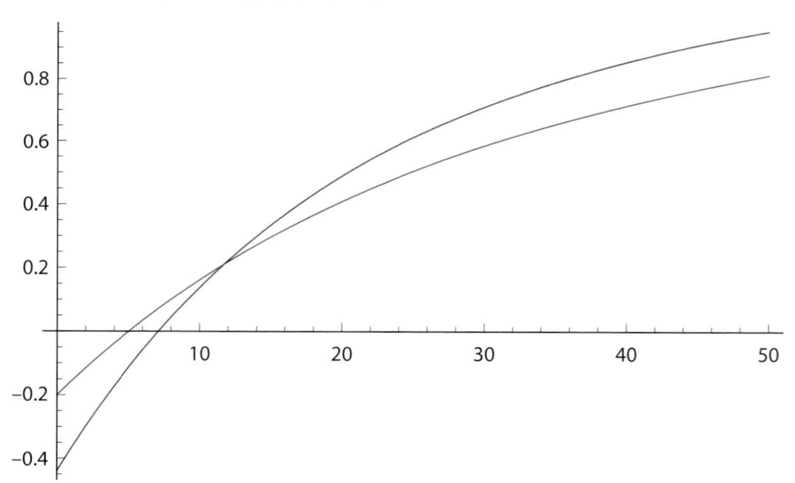

To plot the "phase portrait" of v_1 on v_2 with t as a parameter, we use ParametricPlot.

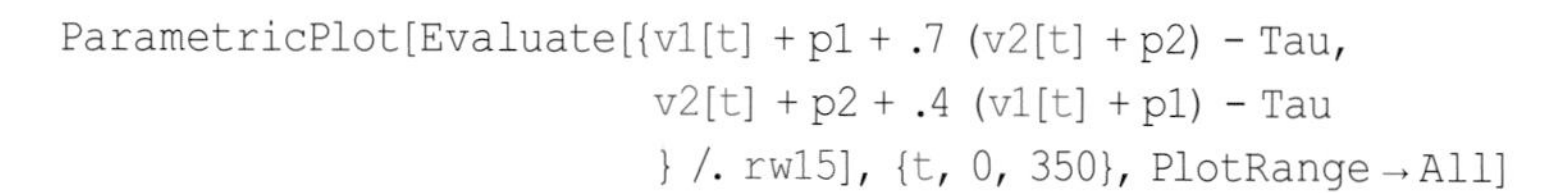

```
ParametricPlot[Evaluate[{v1[t] + p1 + .7 (v2[t] + p2) - Tau,
                         v2[t] + p2 + .4 (v1[t] + p1) - Tau
                         } /. rw15], {t, 0, 350}, PlotRange→All]
```

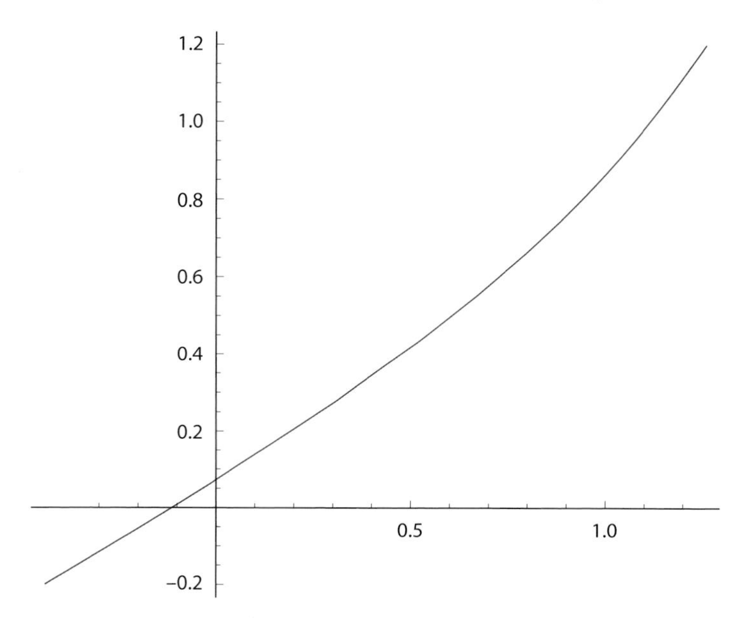

.9 : Goes first

```
Plot[Evaluate[{v1[t] + p1 + .9 (v2[t] + p2) - Tau,
               v2[t] + p2 + .4 (v1[t] + p1) - Tau
               } /. rw15], {t, 0, 50}]
```

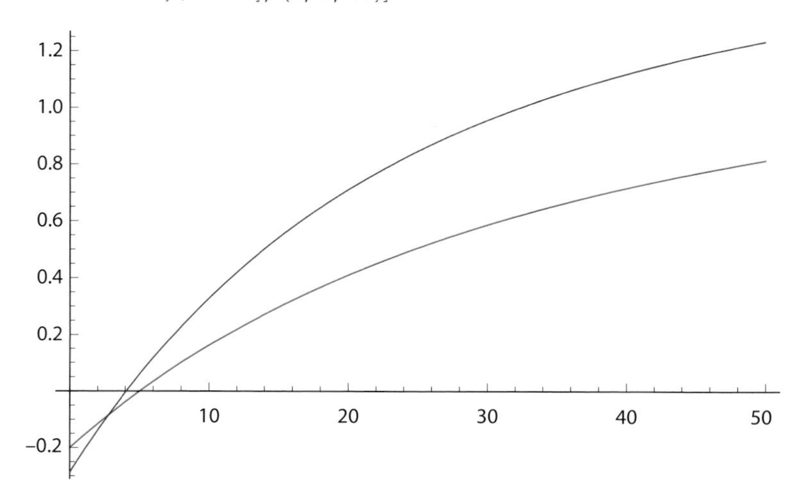

To properly compute dispositional extinction trajectories from $t = 50$, one must first pull out the purely affective levels at that time and initialize the Rescorla-Wagner affective extinction variant at those values, which happen to be 0.86 and 0.60. Then NDSolve is used with $\lambda = 0$ to generate the purely affective extinction curves. Finally, we add the probabilities and subtract the thresholds to yield the full dispositional extinction trajectories given in Figure 19.

```
rw16 = NDSolve[{
              v1'[t] == 0.06 (v1[t]^0) (0 - v1[t]),
              v2'[t] == 0.02 (v2[t]^0) (0 - v2[t]),
              v1[0] == .86, v2[0] == .60},
              {v1, v2}, {t, 650}]
```

```
{{v1 → InterpolatingFunction[{{0., 650.}}, <>], v2 → InterpolatingFunction[{{0., 650.}}, <>]}}
```

```
Plot[Evaluate[{v1[t] + p1 + .9 (v2[t] + p2) - Tau,
              v2[t] + p2 + .4 (v1[t] + p1) - Tau
              } /. rw16], {t, 0, 100}, PlotRange → All]
```

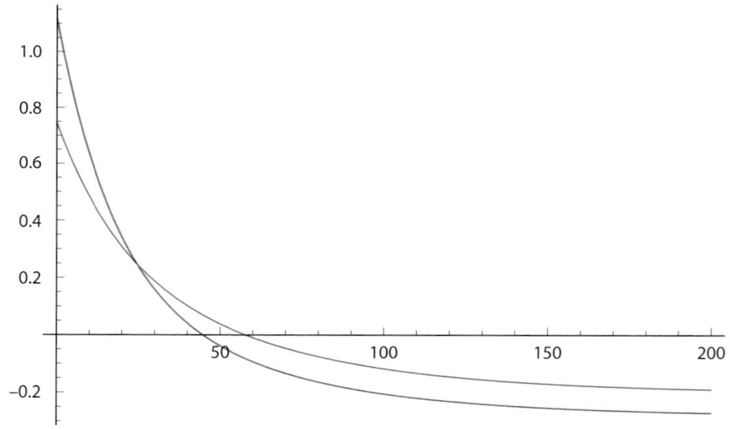

II.3. Three-Agent Runs: Homogeneous Classical Rescorla-Wagner Learners and Heterogeneous Nonclassical (Generalized) Rescorla-Wagner Learners

First, we generate the pair of runs in Figure 25. Agent 3 is the protagonist. In the first run, he assigns zero weight to the other two agents, who are identical (so their red and blue curves coincide, producing a purple curve). With no learning, Agent 3 sits at minus Tau. With maximal learning (weights of 1.0), he goes first. As discussed in the text, three agents is the minimum number required for this phenomenon.

```
p1 = .60
p2 = .60
p3 = .0
Tau = 1.5

rw2 = NDSolve[{
            v1'[t] == .1 (v1[t]^.0) (1 - v1[t]),
            v2'[t] == .1 (v2[t]^.0) (1 - v2[t]),
            v3'[t] == .0 (v3[t]^.0) (1 - v3[t]),
            v1[0] == v2[0] == v3[0] == .0001},
            {v1, v2, v3}, {t, 300}]

Plot[Evaluate[{v1[t] + p1 + .3 (v2[t] + p2) + .0 (v3[t] + p3) - Tau,
            v2[t] + p2 + .3 (v1[t] + p1) + .0 (v3[t] + p3) - Tau,
            v3[t] + p3 + 0 (v2[t] + p2) + 0 (v1[t] + p3) - Tau} /. rw2],
            {t, 0, 30}, PlotRange → All]
```

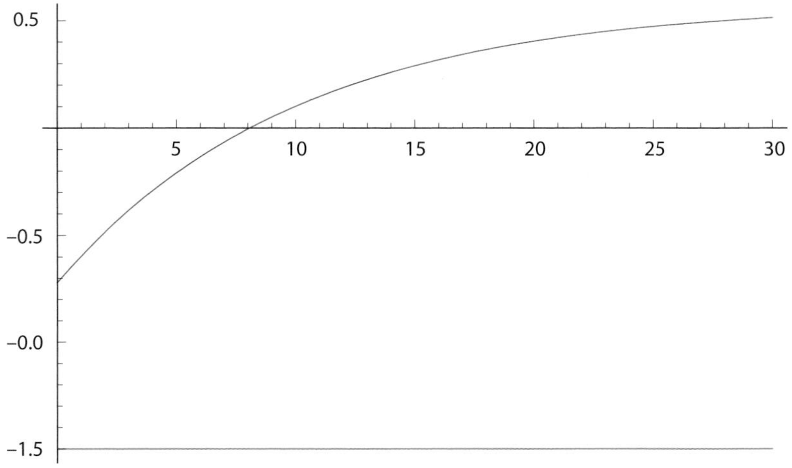

```
Plot[Evaluate[{v1[t] + p1 + .3 (v2[t] + p2) + .0 (v3[t] + p3) - Tau,
                v2[t] + p2 + .3 (v1[t] + p1) + .0 (v3[t] + p3) - Tau,
                v3[t] + p3 + 1 (v2[t] + p2) + 1 (v1[t] + p3) - Tau} /. rw2],
                {t, 0, 30}, PlotRange → All]
```

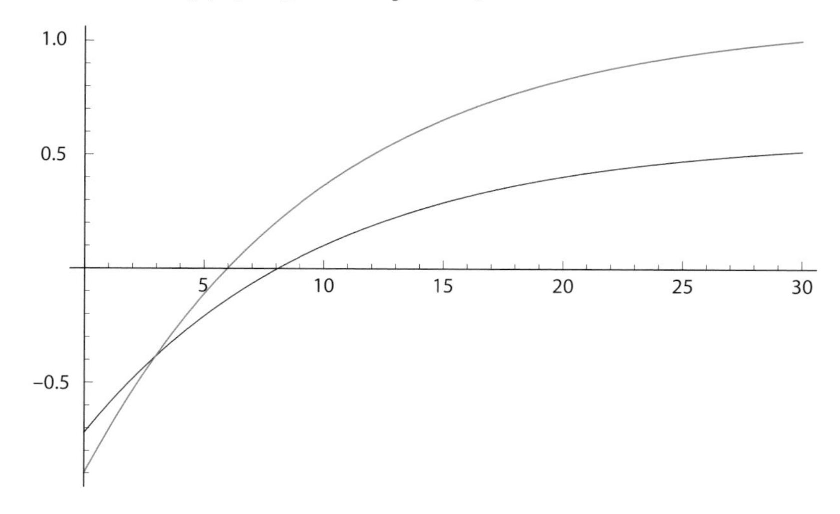

Next, we exploit the heterogeneity afforded by the generalized model, generating the plots of Figure 26. Agent 3 remains a classical learner. But the others have diverse probability estimates and δ exponents, which make them *S*-curve learners. As usual, we first apply NDSolve to the generalized system of differential equations and then show two plots. With zero weights, Agent 3 never acts. With a weight of 0.475, he acts first and ends with the highest disposition.

```
p1 = .80
p2 = .60
p3 = .00
Tau = 1.5
```

```
rw6 = NDSolve[{
            v1'[t] == .10 (v1[t]^1) (1 - v1[t]),
            v2'[t] == .03 (v2[t]^.5) (1 - v2[t]),
            v3'[t] == .02 (v3[t]^.0) (1 - v3[t]),
            v1[0] == v2[0] == v3[0] == .0001},
            {v1, v2, v3}, {t, 300}]*
```

*As per footnote 107, $\delta > 0$ requires positive initial affect. We use 0.0001 here and, purely for maximum comparability, in the preceding case.

```
Plot[Evaluate[{v1[t] + p1 + .3 (v2[t] + p2 + v3[t] + p3) - Tau,
             v2[t] + p2 + .2 (v1[t] + p1 + v3[t] + p3) - Tau,
             v3[t] + p3 + 0 (v1[t] + p1 + v2[t] + p2) - Tau} /. rw6],
             {t, 0, 300}, PlotRange → All]
```

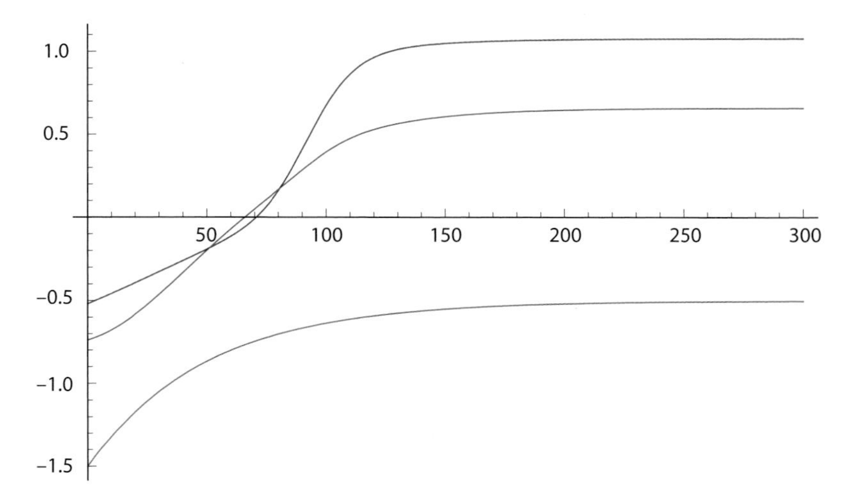


```
Plot[Evaluate[{v1[t] + p1 + .3 (v2[t] + p2 + v3[t] + p3) - Tau,
             v2[t] + p2 + .2 (v1[t] + p1 + v3[t] + p3) - Tau,
             v3[t] + p3 + .475 (v1[t] + p1 + v2[t] + p2) - Tau} /. rw6],
             {t, 0, 300}, PlotRange → All]
```

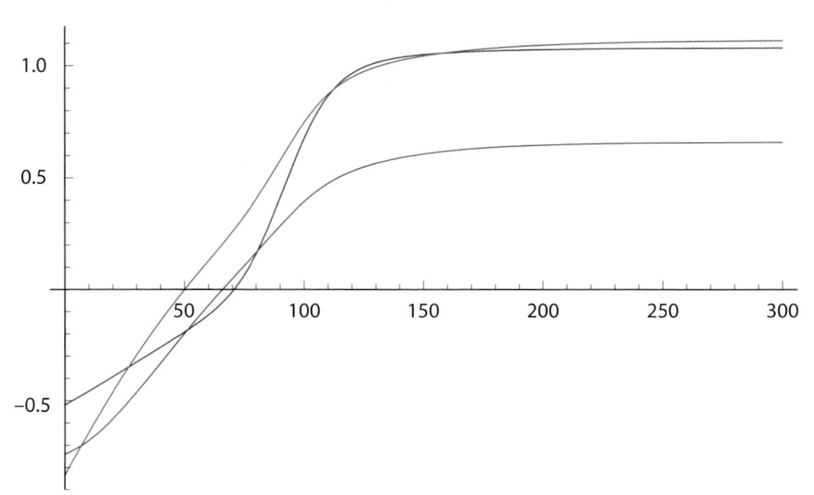

Extinction Phase

Now, we generate the two extinction curves of Figures 32 and 33. In the first, no one has posttraumatic stress and all can reset their λ-values to zero. In the second, Agent 3 can reset only to 0.95, radically delaying her recovery (until $t = 750$), and substantially delaying everyone else's. As before, we first solve the generalized Rescorla-Wagner equations for the purely affective trajectories, initializing at whatever value they attained when extinction begins. In this case, it is clear from the horiziontal net dispositions that the v-trajectories are close to their asymptotic values of 1.0, which we shall use. However, were that not the case, a cute way to extract approximate values, circumventing inspection of the interpolating functions, is to simply plot the affects from 299 to 300, for example, with the following command:

```
Plot[Evaluate[{v3[t], v2[t], v1[t]} /. rw6], {t, 299, 300}]
```

The extinction computations follow:

```
rw7 = NDSolve[{
            v1'[t] == .10 (v1[t]^1) (-v1[t]),
            v2'[t] == .03 (v2[t]^.5) (-v2[t]),
            v3'[t] == .02 (v3[t]^.0) (-v3[t]),
            v1[0] == 1, v2[0] == 1, v3[0] == 1},
            {v1, v2, v3}, {t, 300}]

Plot[Evaluate[{v1[t] + p1 + .3 (v2[t] + p2 + v3[t] + p3) - Tau,
            v2[t] + p2 + .2 (v1[t] + p1 + v3[t] + p3) - Tau,
            v3[t] + p3 + .475 (v2[t] + p2 + v1[t] + p1) - Tau} /. rw7],
            {t, 0, 200}]
```

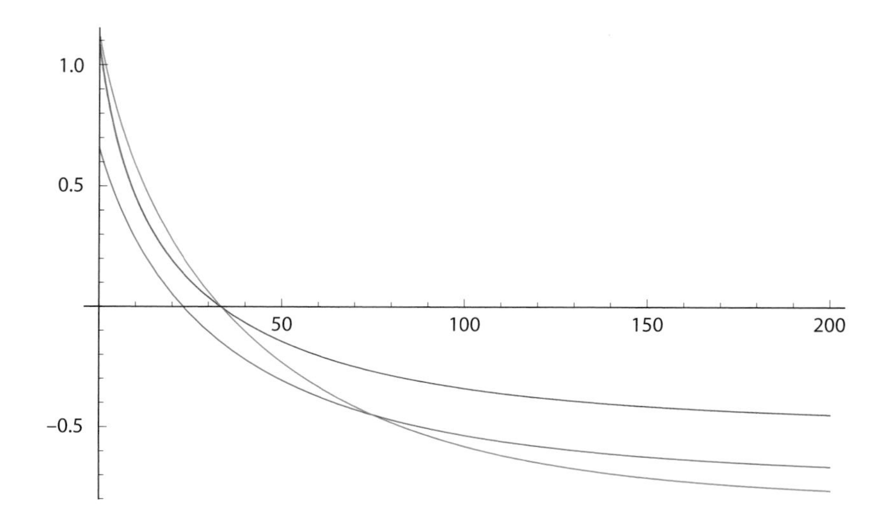

```
rw8 = NDSolve[{
           v1'[t] == .10 (v1[t]^1) (-v1[t]),
           v2'[t] == .03 (v2[t]^.5) (-v2[t]),
           v3'[t] == .02 (v3[t]^.0) (.95 - v3[t]),
           v1[0] == 1, v2[0] == 1, v3[0] == 1},
           {v1, v2, v3}, {t, 3000}]

Plot[Evaluate[{v1[t] + p1 + .3 (v2[t] + p2 + v3[t] + p3) - Tau,
            v2[t] + p2 + .2 (v1[t] + p1 + v3[t] + p3) - Tau,
            v3[t] + p3 + .475 (v2[t] + p2 + v1[t] + p1) - Tau} /. rw8],
            {t, 0, 200}]
```

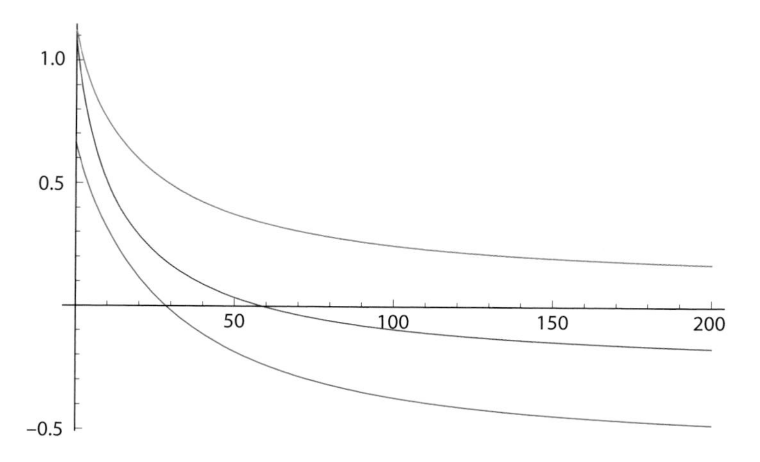

II.4. Group's Disposition Trajectory in a Vector Field

These commands generate the vector field excursion reported in Part I. First, using VectorPlot3D, we generate the negative radial field of Figure 30 and store it as j1.

```
j1 = VectorPlot3D[{-x, -y, -z}, {x, 0, 3}, {y, 0, 3}, {z, 0, 3};
```

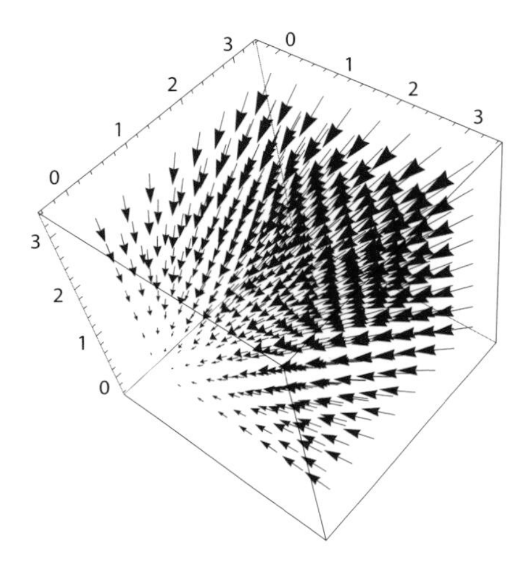

Then, we define the space curve with ParametricPlot3D and, in one move, invoke the Show command to superimpose them, as in Figure 31.

```
j2 = ParametricPlot3D[{t, 2t, 4Sin[t]}, {t, 0, 3}, PlotStyle → {Thick, Red}]
```

```
Show[j1, j2]
```

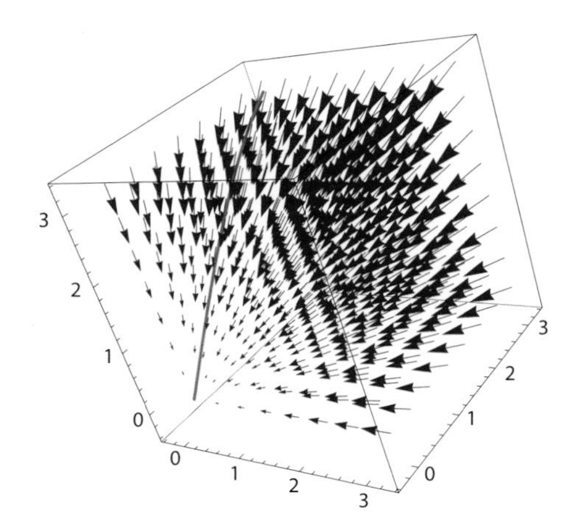

II.5. Strength-Homophily Dynamics

Here, we give Mathematica code for Figures 65, 66, and 67, where all agents are nonclassical with heterogeneous positive δ-values. First, we plot net dispositions and next we plot only the interagent weights over time. In both cases the weight is the sum of affective values times the quantity: 1 minus the absolute value of the affective difference. Third, the dynamic weight vector traces a curve in 3-space.

```
p1 = .80
p2 = .80
p3 = .80
Tau = 1.5

rw9 = NDSolve[{
            v1'[t] == .1 (v1[t]^1.0) (1 - v1[t]),
            v2'[t] == .1 (v2[t]^0.8) (1 - v2[t]),
            v3'[t] == .1 (v3[t]^0.2) (1 - v3[t]),
            v1[0] == v2[0] == v3[0] == .0001},
            {v1, v2, v3}, {t, 600}]

d9 = Plot[Evaluate[{
        v1[t] + p1 + (v1[t] + v2[t]) (1 - Abs[v1[t] - v2[t]])
                (v2[t] + p2) + (v1[t] + v3[t])
                (1 - Abs[v1[t] - v3[t]]) (v3[t] + p3) - Tau,
        v2[t] + p2 + (v1[t] + v2[t]) (1 - Abs[v2[t] - v1[t]])
                (v1[t] + p1) + (v2[t] + v3[t]) (1 - Abs[v2[t] - v3[t]])
                (v3[t] + p3) - Tau,
        v3[t] + p3 + (v2[t] + v3[t]) (1 - Abs[v3[t] - v2[t]])
                (v2[t] + p2) + (v1[t] + v3[t]) (1 - Abs[v3[t] - v1[t]])
                (v1[t] + p3) - Tau} /. rw9], {t, 0, 150}]
```

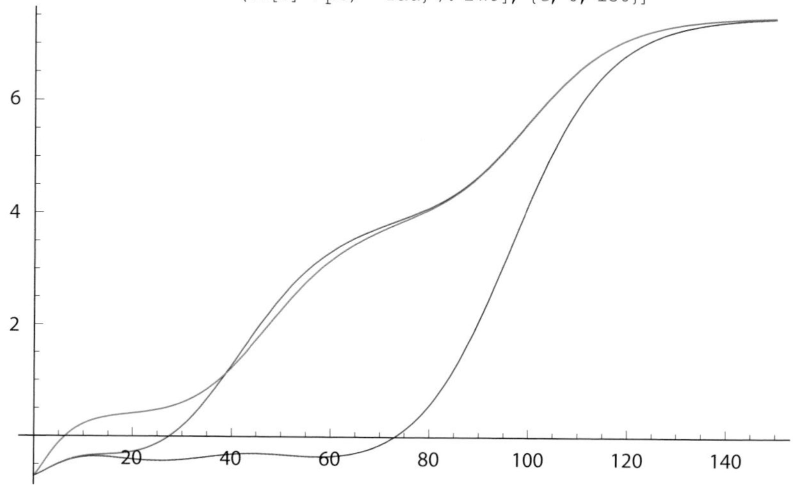

```
w9 = Plot[Evaluate[{v1[t] + v2[t]) (1 - Abs[v1[t] - v2[t]]),
                   (v2[t] + v3[t]) (1 - Abs[v2[t] - v3[t]]),
                   (v1[t] + v3[t]) (1 - Abs[v3[t] - v1[t]])} /. rw9], {t, 0, 150}]
```

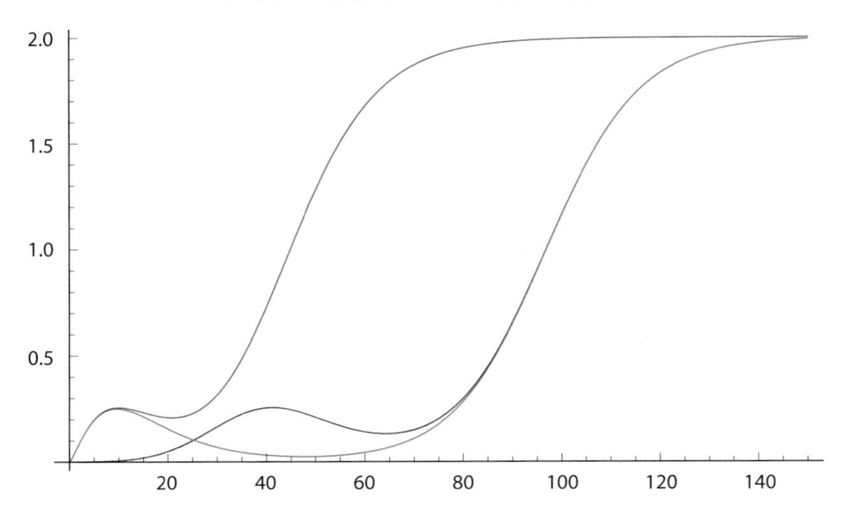


```
ParametricPlot3D[
 Evaluate[{(v1[t] + v2[t]) (1 - Abs[v1[t] - v2[t]]),
           (v2[t] + v3[t]) (1 - Abs[v2[t] - v3[t]]),
           (v1[t] + v3[t]) (1 - Abs[v3[t] - v1[t]])} /. rw9],{t, 0, 150},
 PlotStyle → {Thick, Red}, AxesLabel → {"ω12", "ω23", "ω31"}]
```

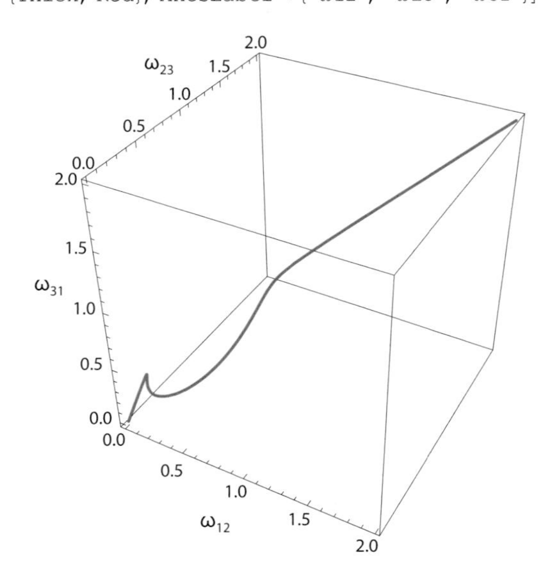

As a final example along these lines, we alter the learning rates of the agents to produce the hysteresis shown in Figure 68.

```
rw10 = NDSolve[{
          v1'[t] == .01 (v1[t]^1.0) (1 - v1[t]),
          v2'[t] == .04 (v2[t]^0.8) (1 - v2[t]),
          v3'[t] == .08 (v3[t]^0.2) (1 - v3[t]),
          v1[0] == v2[0] == v3[0] == .0001},
          {v1, v2, v3}, {t, 1800}]

ParametricPlot3D[Evaluate[
           {(v1[t] + v2[t]) (1 - Abs[v1[t] - v2[t]]),
            (v2[t] + v3[t]) (1 - Abs[v2[t] - v3[t]]),
            (v1[t] + v3[t]) (1 - Abs[v3[t] - v1[t]])} /. rw10],
           {t, 0, 200}, PlotStyle → {Thick, Red},
           AxesLabel → {"ω12", "ω23", "ω31"},
           PlotRange → All]
```

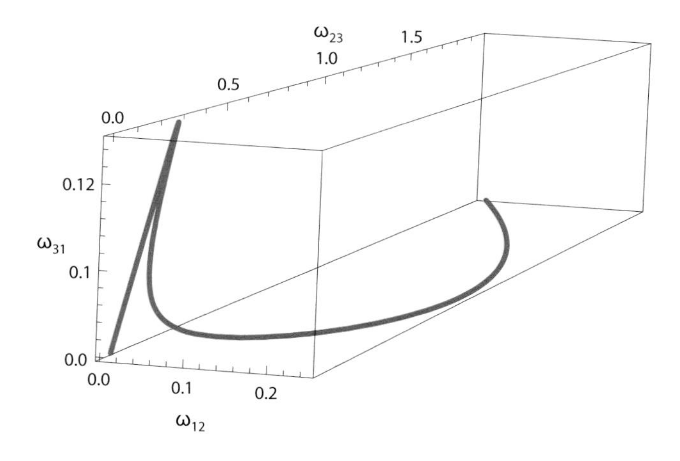

Appendix III.
Agent_Zero NetLogo Source Code

In this appendix, I give the complete *NetLogo* code for Parable One, the Slaughter of Innocents through Dispositional Contagion. Complete *NetLogo* code for every movie run is included in the Applets provided on the book's Princeton University Press Website. These are all interactive, allowing experimentation. All movies are posted separately on the Website. The code starts below.

```
;;;;;;;;;;;;;;;;;;;;;;;;;;;;;
;;    AGENT _ ZERO
;;    Parable 1 Code
;;    Joshua M. Epstein
;;    February 2010
;;    Revised November 2012
;;    Book Sliders [45,0,6,1,1]
;;    Book Seed 1
;;;;;;;;;;;;;;;;;;;;;;;;;;;;;

to use-new-seed                                ; Seed generated by NetLogo
  let my-seed new-seed
  output-print word "Generated seed: " my-seed ; Print it out in command center window
  random-seed my-seed                          ; Use the new seed
end

to use-seed-from-user                          ; Queries user for seed
  let my-seed read-from-string user-input "Enter a random seed (an integer):"
  output-print word "User-entelink-colorseed: " my-seed
  random-seed my-seed
end
```

```
globals[                          ; The more important of these are set w/sliders.
  inactive-color
  active-color
  destroyed-color
  link-color
  deactivation-rate              ; Probablitity a patch returns to inactive from active
  maximum-stopping-time
]

to setup-globals
  set inactive-color yellow  ; Small random added in setup-patches for visualization.
  set active-color orange + 1
  set destroyed-color red - 3
  set deactivation-rate 8
  set link-color red
  set maximum-stopping-time 50000
  end

directed-link-breed [red-links red-link]
links-own [weight]
turtles-own [
            affect
            learning _ rate
            lambda
            delta                ; Generalized Rescorla-Wagner exponent
            threshold
            event_count
            disposition
            probability
            memory
            ]

to setup-links
  ask turtle 0 [ create-red-link-to turtle 1]
              ask red-link 0 1 [
              set color link-color
              set weight 0.3
              ]
  ask turtle 0 [ create-red-link-to turtle 2]
              ask red-link 0 2 [
              set color link-color
              set weight 0.3]
```

```
  ask turtle 1 [ create-red-link-to turtle 0]
               ask red-link 1 0 [
               set color link-color
               set weight 0.3]
  ask turtle 1 [ create-red-link-to turtle 2]
               ask red-link 1 2 [
               set color link-color
               set weight 0.3]
  ask turtle 2 [ create-red-link-to turtle 0]
               ask red-link 2 0 [
               set color link-color
               set weight 0.3]
  ask turtle 2 [ create-red-link-to turtle 1]
               ask red-link 2 1 [
               set color link-color
               set weight 0.3]
end

to setup
  __clear-all-and-reset-ticks
  ;use-new-seed    ; Uncomment to have NetLogo generate new seed for each new run.
  use-seed-from-user ; Base Case seed = 1
  setup-globals
  setup-patches
  setup-turtles
  setup-links
  ;setup-movie     ; Uncomment to make movie
end

to setup-movie
  movie-start "out.mov"
  movie-grab-view        ; Show the initial state
  repeat 750             ; Movie length in frames
   [ go
   movie-grab-view ]
  movie-close
end
```

```
to setup-turtles
  set-default-shape turtles "person"
  create-turtles 3
  ask turtle 0 [
              setxy -9 -9    ; For random positions use random x-cor  random y-cor
              set color blue
              set affect 0.001 ; For delta > 0, if V(0)=0, then V(t)=0 for all t. Hence,
                                 small positive value
              set delta 0
              set lambda 1
              set learning_rate .1
              set threshold .5
              set event_count 0
              set disposition 0
              set probability 0
              set memory []
                repeat memory_length
                  [set memory lput random-float 0 memory]
              ;set memory [0]      ; Recovers memory one original, w/initial estimate
                                     the entry.
              ;set memory [0 0 0 0]
              ;set label who              ; Uncomment to have agent number visisble
              ]

  ask turtle 1 [
              setxy 5 5
              set color blue
              set affect 0.001
              set delta 0
              set lambda 1
              set learning_rate .1
              set threshold .5
              set event_count 0
              set disposition 0
              set probability 0
              ;set memory [ 1 3 5 7 8]
              set memory []
                repeat memory_length
                [set memory lput random-float 0 memory]
              ; show memory
              ; set label who
              ]
```

```
  ask turtle 2 [
               setxy 3 9
               set color blue
               set affect 0.001
               set delta 0
               set lambda 1
               set learning_rate .1
               set threshold .5
               set event_count 0
               set disposition 0
               set probability 0
               set memory []
                  repeat memory_length
                     [set memory lput random-float 0 memory]
               ;set memory [0 0]
               ;set label who
               ]
end

to setup-patches
  ask patches [set pcolor inactive-color + random 2]
end

to go          ; Corresponds to 'main' in C/C++. Placement arbitrary in NetLogo.
  if ticks >= maximum-stopping-time [stop]
  move-turtles
  activate-patches
  update-event_count
  update-affect
  update-probability
  update-disposition
  take-action
  deactivate-patches ; Turns them back to inactive-color from active-color.
  do-plots1
  do-plots2
  do-plots3
  tick
end

to update-event_count
ask turtles[
    if pcolor = active-color [set event_count event_count + 1]
          ]
end
```

```
to update-affect
   ask turtles [
        if pcolor = active-color
           [set affect affect + (learning_rate* (affect ^ delta) * (lambda - affect))]
        if pcolor != active-color
           [set affect affect + (learning_rate* (affect ^ delta) * extinction_rate *(0
              - affect))]
                ]      ; Standard RW extinction for extinction_rate=1.
end

to update-probability
  ask turtles[
  let current_probability
         (count patches in-radius spatial_sample_radius with
         [pcolor = active-color]/(count patches in-radius spatial_sample_radius))
  set memory but-first memory                     ; drops leftmost value
  set memory lput current_probability memory; adds current_prob as rightmost value
  set probability mean memory
  ;set probability median memory      ; Uncomment this, and comment out mean to switch
                                       to moving median.
    ]
  end

to update-disposition   ; Uses Dispostition net of threshold
  ask turtle 0 [
    set disposition affect + probability +
                    [weight] of red-link 1 0 * ([affect] of turtle 1 + [probability] of
                    turtle 1) +
                    [weight] of red-link 2 0 * ([affect] of turtle 2 + [probability] of
                    turtle 2) - threshold]
ask turtle 1 [
    set disposition affect + probability +
                    [weight] of red-link 0 1 * ([affect] of turtle 0 + [probability] of
                    turtle 0) +
                    [weight] of red-link 2 1 * ([affect] of turtle 2 + [probability] of
                    turtle 2) - threshold]
ask turtle 2 [
    set disposition affect + probability +
                    [weight] of red-link 0 2 * ([affect] of turtle 0 + [probability] of
                    turtle 0) +
                    [weight] of red-link 1 2 * ([affect] of turtle 1 + [probability] of
                    turtle 1) - threshold]
end
```

```
to move-turtles
  ask turtles with [who != 0] [  ; This immobilizes Agent_0
    right random 360             ; Adopt random heading
    forward 1
    ]
    end

to take-action
  ask turtles [
    if disposition > 0 [ask patches in-radius action_radius
                        [set pcolor destroyed-color]]
    ]
end

to activate-patches   ; patches in positive quadrant that are not dead
  ask patches with [pxcor > -5 and
                    pycor > -2]
      [if random 1000 < attack_rate and pcolor != destroyed-color
      [set pcolor active-color]]
end

to deactivate-patches
  ask patches with [pxcor > -5 and pycor > -2]
      [if random 100 < deactivation-rate and pcolor != destroyed-color
      [set pcolor inactive-color + random 2]]; they hide again
end

to do-plots1
    set-current-plot "Disposition"
    set-current-plot-pen "turtle 0"
    plot [disposition] of turtle 0
    set-current-plot-pen "turtle 1"
    plot [disposition] of turtle 1
    set-current-plot-pen "turtle 2"
```

```
  plot [disposition] of turtle 2
 end

to do-plots2
  set-current-plot "Probability"
  set-current-plot-pen "turtle 0"
  plot [probability] of turtle 0
  set-current-plot-pen "turtle 1"
  plot [probability] of turtle 1
  set-current-plot-pen "turtle 2"
  plot [probability] of turtle 2
 end

to do-plots3
  set-current-plot "Affect"
  set-current-plot-pen "turtle 0"
  plot [affect] of turtle 0
  set-current-plot-pen "turtle 1"
  plot [affect] of turtle 1
  set-current-plot-pen "turtle 2"
  plot [affect] of turtle 2
 end
```

Appendix IV.
Parameter Settings for Model Runs

Run	Description	Movie	Attack Rate
1	Random movement / no activation	1	0
2	Patch activation only[a]	2	60
3	Parable 1: Slaughter	3	45[c]
4	Parable 2: Agent 0 initiates	4	25
5	Fight	5	30
6	Flight	6	30
7	Darley solo	7	19
8	Darley trio	8	19
9	Dynamic net no growback	9	25
10	Arab Spring	10	6
11	Jury trial	11	Variable[h]
12	Dynamic net structure[j]	12	102
13	Seasonal demand	13	Sinusoidally forced[k]
14	Escalation spiral	14	Endogenous

Note: Some runs are simple variations on these, including alterations in a specific procedure. In these cases, the code variants are given in the text.

a There are a number of ways to arrange this. One is to comment out the move-turtles routine from the "go" procedure and set the threshold to 20, for example, to suppress any reaction.

b In this animation, weights purely determine graphical link thickness.

c Probability of Patch activation: 96/1000 = 0.096. Probability of de-activation is 8/100 = .08

d Heterogeneity across these agent parameters affects the order and timing of agent actions.

e Rather than destroy to radius 2, fight, the action is to move 1 patch in the (–12, –12) direction.

f If Disposition exceeds threshold, move immediately to (–9, –9)

g All initial weights are zero for runs 10–12

h The attack and deactivation rates differ in the pre-trial, courtroom, and jury chamber phases, see code.

i Agents occupy different positions in the three phases.

j Unlike all the other rows, this adds the link_threshold of 0.2.

k Initial attack_rate setting is 300, but activation occurs only if attack_rate • sin (ticks) /10 < a random number < 1000. Deactivation is at 0.08.

Continued

Run	Description	Deactivation Rate
1	Random movement / no activation	[--------------------
2	Patch activation only	8
3	Parable 1: Slaughter	8
4	Parable 2: Agent 0 initiates	8
5	Fight	8
6	Flight	8
7	Darley solo	1
8	Darley trio	1
9	Dynamic net no growback	8
10	Arab Spring	8
11	Jury trial	Variable
12	Dynamic net structure	8
13	Seasonal demand	0.08
14	Escalation spiral	33.3

Run	Description	Link Weights					
		[0, 1]	[0, 2]	[1, 0]	[1, 2]	[2, 0]	[2, 1]
1	Random movement / no activation	0.3	0.3	0.3	0.3	0.3	0.3
2	Patch activation only	0.3[b]	0.3	0.3	0.3	0.3	0.3
3	Parable 1: Slaughter	0.3	0.3	0.3	0.3	0.3	0.3
4	Parable 2: Agent 0 initiates	0.1	0.1	0.7	0.1	0.7	0.1
5	Fight	0.1	0.1	0.1	0.1	0.1	0.1
6	Flight	0.1	0.1	0.1	0.1	0.1	0.1
7	Darley solo	No Links					
8	Darley trio	0.0	0.0	0.1	0.0	0.1	0.0
9	Dynamic net no growback	0.1	0.1	0.1	0.1	0.1	0.1
10	Arab Spring	Endogenous[g] dynamic. See code					
11	Jury trial	Endogenous dynamic. See code					
12	Dynamic net structure	Endogenous dynamic. See code					
13	Seasonal demand	0.3	0.3	0.3	0.3	0.3	0.3
14	Escalation spiral	No Links					

Extinction Rate	Spatial Sample Radius	Action Radius	Memory
--------Irrelevant--------			]
[--------Irrelevant--------			]
0.00	6	1	1
0.00	4	1	1
0.00	4	2	1
0.00	4	1[e]	1
0.00	4	(−9, −9)[f]	1
0.00	4	(−9, −9)	1
0.00	4	1	1
0.00	4	1	1
0.02	30	0	90
0.00	4	1	1
0.10	4	1	1
0.00	4	Endogenous	1

Agent 0					
(x_0, y_0)	V_0	δ	λ	$\alpha\beta$	τ
[--------Irrelevant--------					]
[--------Irrelevant--------				]	20
(−9, −9)	0.001	0.0	1	0.10	0.5
(−9, −9)	0.001	0.0	1	0.10	1.0
(11, 8)	0.001	1.0	1	0.05	0.5
(11, 8)	0.001	1.0	1	0.05	0.5
(7, 7)	0.001	0.0	1	0.05	0.5
(7, 7)	0.001	0.0	1	0.05	0.5
(−9, −9)	0.001	0.0	1	0.05	0.5
(−9, −9)	0.001	0.0	1	0.05	1.5
(−5, −3)[i]	0.001	0.0	1	0.05	0.5
(−9, −9)	0.001	0.0	1	0.05	0.5
(4, 9)	0.001	0.0	1	0.10	1.0
(−9, −9)	0.1	0.0	1	0.20	0.5

Continued

Run	Description	Agent 1					
		(x_0, y_0)	V_0	δ	λ	$\alpha\beta$	τ
1	Random movement / no activation	[-------------------Irrelevant-------------------]					
2	Patch activation only	[---------------Irrelevant---------------]					20
3	Parable 1: Slaughter	(5, 5)	0.001	0.0	1	0.10	0.5
4	Parable 2: Agent 0 initiates	(5, 5)	0.001	0.0	1	0.10	1.0
5	Fight	(5, 5)	0.001	0.5	1	0.03	0.5
6	Flight	(5, 5)	0.001	0.5	1	0.03	0.5
7	Darley solo	[--					
8	Darley trio	(5, 5)	[----Confederates: No activation----]				
9	Dynamic net no growback	(5, 5)	0.001	0.0	1	0.03	0.5
10	Arab Spring	(5, 5)	0.001	0.0	1	0.03	1.5
11	Jury trial	(−7, −7)	0.001	0.0	1	0.035	0.5
12	Dynamic net structure	(5, 5)	0.001	0.0	1	0.03	0.5
13	Seasonal demand	(5, 2)	0.001	0.0	1	0.10	0.85
14	Escalation spiral	(5, 5)	0.1	0.0	1	0.20	0.5

Agent 2						Seed
(x_0, y_0)	V_0	δ	λ	$\alpha\beta$	τ	
[----------Irrelevant----------]						Any
[----------Irrelevant----------]					20	Any
(3, 9)	0.001	0.0	1	0.10	0.5	1
(3, 9)	0.001	0.0	1	0.10	1.0	1
(3, 12)	0.001	0.1	1	0.09	0.5[d]	1
(3, 12)	0.001	0.1	1	0.09	0.5	1
------Absent from run----------]						1
(3, 9)	[----------Confederates: No activation----------]					1
(3, 9)	0.001	0.0	1	0.09	0.5	1
(3, 9)	0.001	0.0	1	0.09	1.5	1
(−10, −4)	0.001	0.0	1	0.02	0.5	1
(3, 9)	0.001	0.0	1	0.09	0.5	2
(11, 12)	0.001	0.0	1	0.10	0.8	1
(3, 9)	0.1	0.0	1	0.20	0.5	2

References

A Letter From The Publisher. (1962). *Time Magazine*.

Adelmann, P. K., and Zajonc, R. B. (1989). Facial efference and the experience of emotion. *Annual Review of Psychology, 40*(1), 249–280.

Adolphs, R., Tranel, D., Damasio, H., and Damasio, A. (1994). Impaired recognition of emotion in facial expressions following bilateral damage to the human amygdala. *Nature, 372*(6507), 669–672.

Adolphs, R., Tranel, D., Hamann, S., Young, A. W., Calder, A. J., Phelps, E. A., Anderson, A., et al. (1999). Recognition of facial emotion in nine individuals with bilateral amygdala damage. *Neuropsychologia, 37*(10), 1111–1117.

Akay, M., Akay, Y. M., and Welkowitz, W. (1994). Automated noninvasive detection of coronary artery disease using wavelet-based neural networks. In *Engineering in Medicine and Biology Society, 1994. Engineering Advances: New Opportunities for Biomedical Engineers. Proceedings of the 16th Annual International Conference of the IEEE*, pp. A12–A13.

Akerlof, G. A., & Shiller, R. J. (2010). *Animal Spirits: How Human Psychology Drives the Economy, and Why It Matters for Global Capitalism*. Princeton: Princeton University Press.

Aristotle (350 BCE; 1958 Edition). *The politics of Aristotle*. E. Barker, Ed. New York: Oxford University Press.

Armony, J. L., Servan-Schreiber, D., Cohen, J. D., and LeDoux, J. E. (1995). An anatomically constrained neural network model of fear conditioning. *Behavioral Neuroscience, 109*(2), 246–257.

Aron, J. L., and Schwartz, I. B. (1984). Seasonality and period-doubling bifurcations in an epidemic model. *Journal of Theoretical Biology, 110*(4), 665–679.

Arthur, W. B. (2013). *Complexity Economics: A Different Framework for Economic Thought*. Oxford: Oxford University Press.

Arzi, A., Shedlesky, L., Ben-Shaul, M., Nasser, K., Oksenberg, A., Hairston, I. S., and Sobel, N. (2012). Humans can learn new information during sleep. *Nature neuroscience, 15*(10), 1460–1465. doi:10.1038/nn.3193

Asch, S. E. (1951). Effects of group pressure upon the modification and distortion of judgments. In H. Guetzkow (Ed.), *Groups, Leadership, and Men* (Vol. 27, pp. 177–190). Pittsburgh: Carnegie Press.

———. (1956). Studies of independence and conformity: A minority of one against a unanimous majority. *Psychological Monographs, 70*(9), 1–70. doi:10.1037/h0093718

Aviezer, H., Trope, Y., and Todorov, A. (2012). Body cues, not facial expressions, discriminate between intense positive and negative emotions. *Science, 338*(6111), 1225–1229. doi:10.1126/science.1224313

Axelrod, R. (1997a). *The Complexity of Cooperation: Agent-based Models of Competition and Collaboration*. Princeton, NJ: Princeton University Press. doi:10.1002/(SICI)1099-0526(199801/02)3:3<46::AID-CPLX6>3.3.CO;2

———. (1997b). The dissemination of culture: A model with local convergence and global polarization. *Journal of Conflict Resolution, 41*(2), 203–226. doi:10.1177/0022002797041002001

Axtell, R. L., Epstein, J. M., & Young, H. P. (1999). The emergence of classes in a multi-agent bargaining model. In H. P. Young & S. Durlauf (Eds.), *Social Dynamics* (pp. 191–211). Cambridge: MIT Press.

Barabasi, A.-L., and Albert, R. (1999). Emergence of scaling in random networks. *Science, 286*(5439), 11.

Baranski, J. V., and Petrusic, W. M. (2010). Aggregating conclusive and inconclusive information: Data and a model based on the assessment of threat. *Journal of Behavioral Decision Making, 23*(4), 383–403. doi:10.1002/bdm.663

Bauer, E. P., Schafe, G. E., and LeDoux, J. E. (2002). NMDA receptors and L-type voltage-gated calcium channels contribute to long-term potentiation and different components of fear memory formation in the lateral amygdala. *Journal of Neuroscience, 22*(12), 5239–5249.

Behrens, T.E.J., Woolrich, M. W., Walton, M. E., and Rushworth, M.F.S. (2007). Learning the value of information in an uncertain world. *Nature Neuroscience, 10*(9), 1214–1221.

Berns, G. S., Chappelow, J., Zink, C. F., Pagnoni, G., Martin-Skurski, M. E., and Richards, J. (2005). Neurobiological correlates of social conformity and independence during mental rotation. *Biological Psychiatry, 58*(3), 245–253.

Berntson, G. G., and Cacioppo, J. T. (Eds.). (2009). *Handbook of Neuroscience for the Behavioral Sciences*. Hoboken, NJ: Wiley. doi:10.1002/9780470478509

Bloom, F., Lazerson, A., and Nelson, C. A. (2001). *Brain, Mind and Behavior* (3rd ed.). New York: Worth Publishers.

Bouton, M. E. (2004). Context and behavioral processes in extinction. *Learning & Memory, 11*(5), 485–494.

Breiter, H. C., Etcoff, N. L., Whalen, P. J., Kennedy, W. A., Rauch, S. L., Buckner, R. L., Strauss, M. M., et al. (1996). Response and habituation of the human amygdala during visual processing of facial expression. *Neuron, 17*(5), 875–887. doi:10.1016/S0896-6273(00)80219-6

Bremner, J. D., Vermetten, E., Schmahl, C., Vaccarino, V., Vythilingam, M., Afzal, N., Grillon, C., et al. (2005). Positron emission tomographic imaging of neural correlates of a fear acquisition and extinction paradigm in women with childhood sexual-abuse-related post-traumatic stress disorder. *Psychological Medicine, 35*(6), 791–806.

Browning, C. R. (1998). *Ordinary Men: Reserve Police Battalion 101 and the Final Solution in Poland*. New York: HarperCollins.

Bush, D.E.A., Schafe, G. E., & LeDoux, J. E. (2009). Neural basis of fear conditioning. In Gary G. Berntson, & John T. Cacioppo (Eds.), *Handbook of Neuroscience for the Behavioral Sciences* (Vol. 2, pp. 762–764). Hoboken, NJ: Wiley.

Butterworth, B. (1999). *The Mathematical Brain*. London: Macmillan.

———. (2005). The development of arithmetical abilities. *The Journal of Child Psychology and Psychiatry and Allied Disciplines, 46*(1), 3–18.

Cacioppo, J. T., Amaral, D. G., Blanchard, J. J., Cameron, J. L., Carter, C. S., Crews, D., Fiske, S., et al. (2007). Social neuroscience: Progress and implications for mental health. *Perspectives on Psychological Science, 2*(2), 99–123. doi:10.1111/j.1745-6916.2007.00032.x

Cacioppo, J. T., and Berntson, G. G. (Eds.). (2009). *Handbook of Neuroscience for the Behavioral Sciences*. Hoboken, NJ: Wiley.

Cacioppo, J. T., and Petty, R. E. (1982). The need for cognition. *Journal of Personality and Social Psychology, 42*(1), 116–131. doi:10.1037/0022-3514.42.1.116

Canetti, E. (1984). *Crowds and Power*. New York: Farrar, Straus and Giroux.

Caramazza, A. (2011). Embodied cognition: Reflecting on mirror neurons. *Annual Review of Neuroscience, 34*.

Carpenter, D., Esterling, K., and Lazer, D. (2003). The strength of strong ties: A model of contact-making in policy networks with evidence from U.S. health politics. *Rationality And Society, 15*(4), 411–440. doi:10.1177/1043463103154001

Centola, D., and Macy, M. (2007). Complex contagions and the weakness of long ties. *American Journal of Sociology, 113*(3), 702–734. doi:10.1086/521848

Cheng, D. T., Knight, D. C., Smith, C. N., Stein, E. A., and Helmstetter, F. J. (2003a). Functional MRI of human amygdala activity during Pavlovian fear conditioning: stimulus processing versus response expression. *Behavioral Neuroscience, 117*(1), 3–10.

———. (2003b). Functional MRI of human amygdala activity during Pavlovian fear conditioning: Stimulus processing versus response expression. *Behavioral Neuroscience, 117*(1), 3–10.

Christakis, N. A., and Fowler, J. H. (2007). The spread of obesity in a large social network over 32 years. *The New England Journal of Medicine, 357*(4), 370–379.

Chomsky, N. (1965). *Aspects of the Theory of Syntax*. Cambridge: MIT Press.

———. (1980). *Rules and Representations*. New York: Columbia University Press.

———. (1986). *Knowledge of Language*. New York: Praeger.

Cook, M., and Mineka, S. (1990). Selective associations in the observational conditioning of fear in rhesus monkeys. *Journal of Experimental Psychology: Animal Behavior Processes 16*, 372–389.

Corbie-Smith, G., Thomas, S. B., and St George, D.M.M. (2002). Distrust, race, and research. *Archives of Internal Medicine, 162*(21), 2458–2463.

Cunningham, W. A., and Brosch, T. (2012). Motivational salience: Amygdala tuning from traits, needs, values, and goals. *Current Directions in Psychological Science, 21*(1), 54–59. doi:10.1177/0963721411430832

Cunningham, W. A, Johnson, M. K., Raye, C. L., Chris Gatenby, J., Gore, J. C., and Banaji, M. R. (2004). Separable neural components in the processing of black and white faces. *Psychological Science, 15*(12), 806–813.

Cunningham, W. A., and Van Bavel, J. J. (2009). A neural analysis of intergroup perception and evaluation. In G. G. Berntson and J. T. Cacioppo (Eds.), *Handbook of Neuroscience for the Behavioral Sciences* (pp. 975–984). Hoboken, NJ: Wiley.

Damasio, A. R. (1994). *Descartes' Error: Emotion, Reason, and the Human Brain*. New York: Avon.

Daniels, R., Taylor, S. C., and Kitano, H. H. (Eds.). (1991). *Japanese Americans: From Relocation to Redress*. Seattle: University of Washington Press.

Danks, D. (2003). Equilibria of the Rescorla–Wagner model. *Journal of Mathematical Psychology, 47*(2), 109–121. doi:10.1016/S0022-2496(02)00016-0

Darley, J. M., and Latané, B. (1968). Bystander intervention in emergencies: diffusion of responsibility. *Journal of Personality and Social Psychology, 8*(4), 377–383.

Darwin, C. (1872). The expression of emotion in man and animals. (P. Ekman, Ed.). University of Chicago Press. Retrieved from ftp://uiarchive.cso.uiuc.edu/pub/etext/gutenberg//etext/pg/etext98/eemaa10.zip.

Davidson, R. J., Scherer, K. R., and Goldsmith, H. H. (Eds.). (2003). *Handbook of the Affective Sciences*. Oxford: Oxford University Press.

Davis, M., and Whalen, P. J. (2001). The amygdala: vigilance and emotion. *Molecular Psychiatry*, *6*(1), 13–34.

Dawes, R. M. (1999). A message from psychologists to economists: Mere predictability doesn't matter like it should (without a good story appended to it). *Journal of Economic Behavior and Organization*, *39*(1), 29–40. doi:10.1016/S0167-2681(99)00024-4

———. (2001). *Everyday Irrationality: How Pseudo-scientists, Lunatics, and the Rest of Us Systematically Fail to Think Rationally*. Boulder: Westview Press.

Dawes, R. M., and Corrigan, B. (1974). Linear models in decision making. *Psychological Bulletin*, *81*(2), 95–106. doi:10.1037/h0037613

Deese, J., and Kaufman, R. (1957). Serial effects in recall of unorganized and sequentially organized verbal material. *Journal of Experimental Psychology*, *54*, 180–187.

De Gelder, B., Snyder, J., Greve, D., Gerard, G., and Hadjikhani, N. (2004). Fear fosters flight: A mechanism for fear contagion when perceiving emotion expressed by a whole body. *Proceedings of the National Academy of Sciences*, *101*(47), 16701–6. doi:10.1073/pnas.0407042101

De Groot, J.H.B., Smeets, M.A.M., Kaldewaij, A., Duijndam, M.J.A., and Semin, G. R. (2012). Chemosignals communicate human emotions. *Psychological Science*, *23*(11), 1417–24. doi:10.1177/0956797612445317

Delgado, M. R., Nearing, K. I., Ledoux, J. E., and Phelps, E. A. (2008). Neural circuitry underlying the regulation of conditioned fear and its relation to extinction. *Neuron*, *59*(5), 829–838.

Descartes, R. J. Epstein (Ed.). (1965). A discourse on method, and other works. Translated by E. E. Haldane and G.R.T. Ross. Abridged, edited, and with an introduction by Joseph Epstein. New York: Washington Square Press.

DeSteno, D., Petty, R. E., Wegener, D. T., & Rucker, D. D. (2000). Beyond valence in the perception of likelihood: the role of emotion specificity. *Journal of Personality and Social Psychology*, *78*(3), 397.

Doherty, R. W. (1998). Emotional contagion and social judgment. *Motivation and Emotion*, *22*(3), 187–209. doi:10.1023/A:1022368805803

Earn, D. J., Rohani, P., Bolker, B. M., and Grenfell, B. T. (2000). A simple model for complex dynamical transitions in epidemics. *Science*, *287*(5453), 667–670.

Eaton, W. W. (2001). *The Sociology of Mental Disorders* (3rd ed.). New York: Praeger.

Edman, Irwin (ed.) (1956). The Philosophy of Plato the Jowett Translation. 181 Edition. New York: Modern Library.

Ehrenstein Illusion. (n.d.). *New World Encyclopedia*. Retrieved from http://www.newworldencyclopedia.org/entry/Ehrenstein_illusion.

Eichenbaum, H. (2000). A cortical-hippocampal system for declarative memory. *Nature Reviews Neuroscience*, *1*(1), 41–50. doi:10.1038/35036213

———. (2012). *The Cognitive Neuroscience of Memory: An Introduction* (2nd ed.). Oxford: Oxford University Press.

Ellner, S. P., Bailey, B. A., Bobashev, G. V., Gallant, A. R., Grenfell, B. T., and Nychka, D. W. (1998). Noise and nonlinearity in measles epidemics: combining mechanistic and statistical approaches to population modeling. *The American Naturalist*, *151*(5), 425–440.

Epley, N., and Gilovich, T. (1999). Just going along: Nonconscious priming and conformity to social pressure. *Journal of Experimental Social Psychology, 35*(6), 578–589. doi:10.1006/jesp.1999.1390

Epstein, J. M. (1997). *Nonlinear Dynamics, Mathematical Biology, and Social Science.* Reading, MA: Addison-Wesley.

———. (1998). Zones of cooperation in demographic prisoner's dilemma. *Complexity, 4*(2), 36–48. doi:10.1002/cplx.10092

———. (2002). Modeling civil violence: An agent-based computational approach. *Proceedings of the National Academy of Sciences of the United States of America, 99* (Suppl. 3), 7243–7250.

———. (2006). *Generative Social Science: Studies in Agent-based Computational Modeling.* Princeton, NJ: Princeton University Press.

———. (2008, October 31). Why model? *Journal of Artificial Societies and Social Simulation 11*(4), 12.

———. (2009). Modelling to contain pandemics. *Nature, 460*(7256), 687. doi:10.1038/460687a

Epstein, J. M, and Axtell, R. (1996). *Growing Artificial Societies: Social Science from the Bottom Up.* Cambridge: MIT Press (A Bradford Book) and Washington, DC: Brookings Institution Press.

Epstein, J. M, Goedecke, D. M., Yu, F., Morris, R. J., Wagener, D. K., and Bobashev, G. V. (2007). Controlling pandemic flu: The value of international air travel restrictions, *PLoS ONE, 2*(5), 11.

Epstein, J. M, Pankajakshan, R., and Hammond, R. A. (2011). Combining computational fluid dynamics and agent-based modeling: A new approach to evacuation planning, *PLoS ONE, 6*(5), 5.

Epstein, J. M, Parker, J., Cummings, D., and Hammond, R. A. (2008). Coupled contagion dynamics of fear and disease: mathematical and computational explorations, *PLoS ONE, 3*(12), 11.

Epstein, S. D. 2000. *Essays in Syntactic Theory.* London: Routledge. doi:10.4324/9780203454701

Epstein, S. D, and Hornstein, N. (Eds.). (1999). *Working Minimalism.* Cambridge: MIT Press.

Epstein, S. D., and Seely, T. D. (Eds.). (2002). *Derivation and Explanation in the Minimalist Program.* Oxford: Blackwell Publishers.

Festinger, L. (1957). *A Theory of Cognitive Dissonance.* Evanston, IL: Row, Peterson.

Fischhoff, B. (2005). The psychological perception of risk. In D. Kamien (Ed.), *Handbook of Terrorism and Counter-terrorism* (pp. 463–492). New York: McGraw-Hill.

Fischhoff, B., Slovic, P., and Lichtenstein, S. (1979). Subjective sensitivity analysis. *Organizational Behavior and Human Performance, 23*(3), 339–359. doi:10.1016/0030-5073(79)90002-3

Fitzhugh, R. (1961). Impulses and physiological states in theoretical models of nerve membrane. *Biophysical Journal, 1*(6), 445–466.

Fodor, J. A. (1983). *Modularity of Mind.* In E. N. Zalta (Ed.), *The Stanford Encyclopedia of Philosophy* (pp. 558–560). Cambridge: MIT Press.

Frangeul, F. (2011). D'où vient la "révolution du jasmin"? *Europe1.* Retrieved from http://www.europe1.fr/International/D-ou-vient-la-revolution-du-jasmin-375743/.

Freeman, J. A., and Skapura, D. M. (1991). *Neural Networks: Algorithms, Applications, and Programming Techniques*. Reading, MA: Addison-Wesley.

Freimuth, V. S., Quinn, S. C., Thomas, S. B., Cole, G., Zook, E., and Duncan, T. (2001). African Americans' views on research and the Tuskegee Syphilis Study. *Social Science Medicine*, *52*(5), 797–808.

French, A. P. (1971). *Vibrations and Waves*. New York: W. W. Norton.

Funayama, E. S., Grillon, C., Davis, M., and Phelps, E. A. (2001). A double dissociation in the affective modulation of startle in humans: effects of unilateral temporal lobectomy. *Journal of Cognitive Neuroscience*, *13*(6), 721–729.

Gazzola, V., Aziz-Zadeh, L., and Keysers, C. (2006). Empathy and the somatotopic auditory mirror system in humans. *Current Biology*, *16*(18), 1824–1829.

Gazzola, V., Van der Worp, H., Mulder, T., Wicker, B., Rizzolatti, G., and Keysers, C. (2007). Aplasics born without hands mirror the goal of hand actions with their feet. *Current Biology*, *17*(14), 1235–40. doi:10.1016/j.cub.2007.06.045

Gelman, R., and Butterworth, B. (2005). Number and language: how are they related? *Trends in Cognitive Sciences*, *9*(1), 6–10.

Gigerenzer, G., and Gaissmaier, W. (2011). Heuristic decision making. *Annual Review of Psychology*, *62*(1), 451–482.

Gilovich, T. (1981). Seeing the past in the present: The effect of associations to familiar events on judgments and decisions. *Journal of Personality and Social Psychology*, *40*(5), 797–808. doi:10.1037//0022-3514.40.5.797

Gilovich, T., Griffin, D., and Kahneman, D. (Eds.) (2002). *Heuristics and Biases: The Psychology of Intuitive Judgment*. Cambridge: Cambridge University Press.

Gilovich, T., Savitsky, K., and Medvec, V. H. (1998). The illusion of transparency: Biased assessments of others' ability to read one's emotional states. *Journal of Personality and Social Psychology*, *75*(2), 332–346.

Glimcher, P. W. (2011). Understanding dopamine and reinforcement learning: the dopamine reward prediction error hypothesis. *Proceedings of the National Academy of Sciences of the United States of America*, *108*(Supplement 3), 15647–15654. doi:10.1073/pnas.1014269108

Glimcher, P. W., Fehr, E., Camerer, C., and Poldrack, R. A. (Eds.). (2008). *Neuroeconomics: Decision Making and the Brain*. London: Academic Press.

Gloor, P. (1992). Role of the amygdala in temporal lobe epilepsy. In J. P. Aggleton (Ed.), *The Amygdala: Neurobiological Aspects of Emotion, Memory, and Mental Dysfunction* (pp. 505–538). New York: Wiley-Liss.

Goswami, S., Cascardi, M., Rodríguez-Sierra, O. E., Duvarci, S., & Paré, D. (2010). Impact of predatory threat on fear extinction in Lewis rats. *Learning & Memory*, *17*(10), 494–501.

Granovetter, M. (1978). Threshold models of collective behavior. *American Journal of Sociology*, *83*(6), 1420–1443. doi:10.1086/226707

———. (1983). The strength of weak ties: A network theory revisited, *Sociological Theory*, *1*(1), 201–233. doi:10.2307/202051

Guckenheimer, J., and Holmes, P. (1983). *Nonlinear Oscillations, Dynamical Systems, and Bifurcations of Vector Fields*. New York: Springer.

Haegerstrom-Portnoy, G., Schneck, M. E., and Brabyn, J. A. (1999). Seeing into old age: Vision function beyond acuity. *Optometry and Vision Science:* Official publication of *The American Academy of Optometry*, *76*(3), 141–58.

Hale, J. K., & Koçak, H. (1991). *Dynamics and Bifurcations*. New York: Springer-Verlag.

Hammond, R. A, and Axelrod, R. (2006a). Evolution of contingent altruism when cooperation is expensive. *Theoretical Population Biology*, *69*(3), 333–338.

———. (2006b). The evolution of ethnocentrism. *Journal of Conflict Resolution*, *50*(6), 926–936. doi:10.1177/0022002706293470

Hammond, R. A., and Epstein, J. M. (2007). *Exploring Price-Independent Mechanisms in the Obesity Epidemic*. Center on Social and Economic Dynamics Working Paper No. 48. Washington, DC: Brookings Institution.

Harms, C. A., Cooper, D., and Tanaka, H. (2011). Exercise physiology of normal development, sex differences, and aging. *Comprehensive Physiology*, *1*(4), 1649–1678.

Harris, B. (1979). Whatever happened to little Albert? *American Psychologist*, *34*(2), 151–160. doi:10.1037//0003-066X.34.2.151

Hart, A. J., Whalen, P. J., Shin, L. M., McInerney, S. C., Fischer, H., and Rauch, S. L. (2000). Differential response in the human amygdala to racial outgroup vs. ingroup face stimuli. *NeuroReport*, *11*(11), 2351–2355.

Hastie, R. (Ed.). (1993). *Inside the Juror: The Psychology of Juror Decision Making*. Cambridge: Cambridge University Press.

Hastie, R., Penrod, S. D., and Pennington, N. (1983). *Inside the Jury*. Cambridge: Havard University Press.

Hatfield, E., Cacioppo, J. T., and Rapson, R. L. (1994). Emotional contagion. New York: Cambridge University Press.

Haviland-Jones, J. M., and Wilson, P. J. (2008). A "nose" for emotion: Emotional information and challenges in odors and semiochemicals . In M. Lewis, J. M. Haviland-Jones, and L. F. Barrett (Eds.), *Handbook of Emotions* (3rd ed., pp. 235–248). New York: The Guilford Press.

Hebb, D. O. (1949). *The Organization of Behavior*. New York: Wiley.

Higbee, K. L. (1969). Fifteen years of fear arousal: Research on threat appeals: 1953–1968. *Psychological Bulletin*, *72*(6), 426–444.

Hobbes, T. (1651; 1958 Edition). *Leviathan*. H.W. Schneider (Ed.). New York: Bobbs-Merrill.

Hodgkin, A. L., and Huxley, A. F. (1952). A quantitative description of membrane current and its application to conduction and excitation in nerve. *The Journal of Physiology*, *117*(4), 500–544.

Hofstadter, R. (1964). The paranoid style in American politics and other essays. *The American Historical Review*, *72*(1), 281. doi:10.2307/1848349

Holland, P., and Gallagher, M. (1999). Amygdala circuitry in attentional and representational processes. *Trends in Cognitive Sciences*, *3*(2), 65–73. doi:10.1016/S1364-6613 (98)01271-6

Hume, D. (1739; 2000 Edition). *A Treatise of Human Nature*. D. F. Norton and M. J. Norton (Eds.). Oxford: Oxford University Press.

———. (1748; 2008 Edition). *An Enquiry Concerning Human Understanding*. P. F. Millican (Ed.). Oxford: Oxford University Press.

Iacoboni, M. (2009). Imitation, empathy, and mirror neurons. *Annual Review of Psychology*, *60*(September 2008), 653–670.

Jackson, E. A. (1992). *Perspectives of Nonlinear Dynamics*. Two Volumes (2nd ed.). Cambridge: Cambridge University Press.

James, W. (1884). What is an emotion? *Mind*, *9*(34), 188–205. doi:10.1093/mind/os-IX .34.188

Janis, I. L., and Feshbach, S. (1953). Effects of fear-arousing communications. *The Journal of Abnormal and Social Psychology*, *48*(1), 78–92. doi:10.1037/h0060732

Jihad. (2013). *BBC Religions: Islam*. Retrieved from http://www.bbc.co.uk/religion/religions/islam/beliefs/jihad_1.shtml.

Joseph, R. (1996). *Neuropsychiatry, Neuropsychology, and Clinical Neuroscience: Emotion, Evolution, Cognition, Language, Memory, Brain Damage, and Abnormal Behavior* (2nd ed.). Baltimore: Williams & Wilkins.

Kahneman, D. (2003). A perspective on judgment and choice: Mapping bounded rationality. *American Psychologist*, *58*(9), 697–720. doi:10.1037/0003-066X.58.9.697

———. (2011). *Thinking, Fast and Slow*. New York: Farrar, Straus and Giroux.

Kahneman, D., and Frederick, S. (2002). Representativeness revisited: Attribute substitution in intuitive judgment. In Thomas Gilovich, Dale Griffin, and Daniel Kahneman, eds, *Heuristics and Biases: The Psychology of Intuitive Judgment* (pp. 49–81). New York: Cambridge University Press.

Kahneman, D., and Tversky, A. (1972). Subjective probability: A judgment of representativeness. *Cognitive Psychology*, *3*(3), 430–454. doi:10.1016/0010-0285(72)90016-3

———. (1973). On the psychology of prediction. *Psychological Review*, *80*(4), 237–251. doi: 10.1037/h0034747

———. (1996). On the reality of cognitive illusions. *Psychological Review*, *103*(3), 582–591.

Kaldor, N. (1967). *Strategic Factors in Economic Development*. Ithaca: Cornell University Press.

Kamin, L. J. (1969). Predictability, surprise, attention, and conditioning. In B. A. Campbell and R. M. Church (Eds.), *Punishment and Aversive Behavior* (pp. 279–296). New York: Appleton-Century-Crofts.

Kandel, E. R. (2001). The molecular biology of memory storage: A dialogue between genes and synapses. *Science*, *294*(5544), 1030–1038. doi:10.1126/science.1067020

Kanizsa, G. (1955). Margini quasi-percettivi in campi con stimolazione omogenea. *Rivista di Psicologia*, *49*(1), 7.

Kanwisher, N. (2010). Functional specificity in the human brain: A window into the functional architecture of the mind. *Proceedings of the National Academy of Sciences*, *107*(25), 11163–11170. doi:10.1073/pnas.1005062107

Kermack, W. O., and McKendrick, A. G. (1927). A contribution to the mathematical theory of epidemics. *Proceedings of the Royal Society of London. Series A*, *115*(772), 700–721. doi:10.1098/rspa.1927.0118

Kindleberger, C. P. (1978; 2000 Edition). *Manias, Panics and Crashes* (4th ed.). New York: Wiley. doi:10.1057/9780230628045

Klüver, H., and Bucy, P. C. (1937). "Psychic blindness" and other symptoms following bilateral temporal lobectomy in Rhesus monkeys. *American Journal of Physiology*, *119*, 352–353.

Knierim, J. J. (2009). Imagining the possibilities: Ripples, routes, and reactivation. *Neuron*, *63*(4), 421–423.

Knierim, J. J., and McNaughton, B. L. (2001). Hippocampal place-cell firing during movement in three-dimensional space. *Journal of Neurophysiology*, *85*(1), 105–116.

Koelle, K., Pascual, M., and Yunus, M. (2005). Pathogen adaptation to seasonal forcing and climate change. *Proceedings. Biological sciences/The Royal Society*, *272*(1566), 971–977. doi:10.1098/rspb.2004.3043

Koriat, A., Lichtenstein, S., and Fischhoff, B. (1980). Reasons for confidence. *Journal of Experimental Psychology: Human Learning and Memory*, *6*(2), 107–118. doi:10.1037/0278-7393.6.2.107

Kremer, M. (1996). Integrating behavioral choice into epidemiological models of AIDS. *The Quarterly Journal of Economics*, *111*(2), 549–573. doi:10.2307/2946687

Kross, E., Berman, M. G., Mischel, W., Smith, E. E., and Wager, T. D. (2011). Social rejection shares somatosensory representations with physical pain. *Proceedings of the National Academy of Sciences of the United States of America*, *108*(15), 6270–6275. doi:10.1073/pnas.1102693108

Krueger, L. E. (1972). Perceived numerosity. *Perception and Psychophysics*, *11*(1), 5–9. doi:10.3758/BF03212674

———. (1982). Single judgments of numerosity. *Perception and Psychophysics*, *31*(2), 175–182. doi:10.3758/BF03206218

LaBar, K. S., Gatenby, J. C., Gore, J. C., and Phelps, E. A. (1998). Role of the amygdala in emotional picture evaluation as revealed by fMRI. *Journal of Cognitive Neuroscience, Suppl. S.*

Laird, J. E., Newell, A., and Rosenbloom, P. S. (1987). SOAR: An architecture for general intelligence. *Artificial Intelligence*, *33*(1), 1–64. doi:10.1016/0004-3702(87)90050-6

Lasker R. (2004). *Redefining Readiness: Terrorism Planning Through the Eyes of the Public.* New York: New York Academy of Medicine.

Latané, B. (1981). The psychology of social impact. *American Psychologist*, *36*(4), 343–356.

Latané, B., and Darley, J. M. (1968). Group inhibition of bystander intervention in emergencies. *Journal of Personality and Social Psychology*, *10*(3), 215–221.

Lazer, D. (2001). The co-evolution of individual and network. *The Journal of Mathematical Sociology*, *25*(1), 69–108. doi:10.1080/0022250X.2001.9990245

Le Bon, G. (1895; 2001 Edition). *The Crowd: A Study of the Popular Mind.* Mineola, NY: Dover Publications.

LeDoux, J. E. (1994). Emotion, memory and the brain. *Scientific American*, *270*(6), 50–57.

———. (1996). *The Emotional Brain: The Mysterious Underpinnings of Emotional Life.* New York: Simon and Schuster.

———. (2002). *Synaptic Self.* New York: Viking.

———. (2003). The emotional brain, fear, and the amygdala. *Cellular and Molecular Neurobiology*, *23*(4-5), 727–738.

———. (2007). The amygdala. *Current Biology*, *17*(20), R868–R874.

———. (2008). Amygdala. *Scholarpedia*, *3*(4), 2698. doi:10.4249/scholarpedia.2698

———. (2009). Emotion systems and the brain. In L. R. Squire (Ed.), *Encyclopedia of Neuroscience* (pp. 903–908). Oxford: Academic Press.

———. (2012). Rethinking the emotional brain. *Neuron*, *73*(4), 653–676. doi:10.1016/j.neuron.2012.02.004

Ledoux, J. E., and Phelps, E. A. (2008). Emotional networks in the brain. In Michael Lewis, J. M. Haviland-Jones, and L. F. Barrett (Eds.), *Handbook of Emotions.* New York: The Guilford Press.

Lee, R. F., Dai, W., and Jones, J. (2012). Decoupled circular-polarized dual-head volume coil pair for studying two interacting human brains with dyadic fMRI. *Magnetic Resonance in Medicine: Official Journal of the Society of Magnetic Resonance in Medicine/Society of Magnetic Resonance in Medicine*, *68*(4), 1087–96. doi:10.1002/mrm.23313

Lerner, J. S., Gonzalez, R. M., Small, D. A., and Fischhoff, B. (2003). Effects of fear and anger on perceived risks of terrorism: A national field experiment. *Psychological Science, 14*(2), 144–150.

Levitan, L. C., and Visser, P. S. (2009). Social network composition and attitude strength: Exploring the dynamics within newly formed social networks. *Journal of Experimental Social Psychology, 45*(5), 1057–1067. doi:10.1016/j.jesp.2009.06.001

Lewis, Michael, Haviland-Jones, J. M., and Barrett, L. F. (2008). *Handbook of Emotions* (3rd ed.). New York: Guilford Press.

Lichtenstein, S., Fischhoff, B., and Phillips, L. D. (1982). Calibration of probabilities: The state of the art to 1980. In D. Kahneman, P. Slovic, and A. Tversky (Eds.), *Judgment under Uncertainty: Heuristics and Biases*. Cambridge: Cambridge University Press.

Lichtenstein, S., Slovic, P., Fischhoff, B., Layman, M., and Combs, B. (1978). Judged frequency of lethal events. *Journal of Experimental Psychology: Human Learning and Memory, 4*(6), 551–578. doi:10.1037/0278-7393.4.6.551

Lindquist, K. A., Wager, T. D., Kober, H., Bliss-Moreau, E., & Barrett, L. F. (2012). The brain basis of emotion: A meta-analytic review. *Behavioral and Brain Sciences*, 35, 121–143.

Loewenstein, G. F., Weber, E. U., Hsee, C. K., & Welch, N. (2001). Risk as feelings. *Psychological Bulletin, 127*(2), 267.

Lotka, A. J. (1910). Contribution to the theory of periodic reaction. *The Journal of Physical Chemistry A, 14*(3), 271–274.

Loughman, J., Davison, P. A., Nolan, J. M., Akkali, M. C., and Beatty, S. (2010). Macular pigment and its contribution to visual performance and experience . *Journal of Optometry, 3*, 74–90.

Lyons, R. (2011). The spread of evidence-poor medicine via flawed social-network analysis. *Statistics Politics and Policy, 2*(1), 1–16.

Mackay, C. (1852). *Extraordinary Popular Delusions and the Madness of Crowds*. London: Office of the National Illustrated Library. Radford Edition (2008).

Marsden, J. E., and Tromba, A. (2011). *Vector Calculus* (6th ed.). New York: Freeman.

Marshall, R. D., Bryant, R. A., Amsel, L., Suh, E. J., Cook, J. M., and Neria, Y. (2007). The psychology of ongoing threat: Relative risk appraisal, the September 11 attacks, and terrorism-related fears. *American Psychologist, 62*(4), 304–316.

Martin, M., and Koppel, T. (2008). The "Last Lynching": How Far Have We Come? *Tell Me More*, NPR. Retrieved from http://www.npr.org/templates/story/story.php?storyId=95672737.

Marx, K. (1869; 2013 Edition). *The Eighteenth Brumaire of Louis Bonaparte*. The Project Gutenberg. EBook 134. Retrieved from http://www.gutenberg.org/files/1346/1346-h/1346-h.htm.

Mayford, M., Mansuy, I. M., Muller, R. U., and Kandel, E. R. (1997). Memory and behavior: A second generation of genetically modified mice. *Current Biology, 7*(9), R580–R589.

Milgram, S. (1963). Behavioral study of obedience. *Journal of Abnormal Psychology, 67*(4), 371–378.

Miller, J. H., and Page, S. E. (2010). *Complex Adaptive Systems: An Introduction to Computational Models of Social Life*. Princeton, NJ: Princeton University Press. doi:10.1016/S1460-1567(08)10011-3

Mineka, S., and Cook, M. (1993). Mechanisms involved in the observational conditioning of fear. *Journal of Experimental Psychology: General, 122*(1), 23–38.

Mineka, S., Davidson, M., Cook, M., and Keir, R. (1984). Observational conditioning of snake fear in rhesus monkeys. *Journal of Abnormal Psychology, 93*(4), 355–372. doi:10.1037/0021-843X.93.4.355

Mischel, W., Ayduk, O., Berman, M. G., Casey, B. J., Gotlib, I. H., Jonides, J., Kross, E., et al. (2011). "Willpower" over the life span: Decomposing self-regulation. *Social Cognitive and Affective Neuroscience*, *6*(2), 252–256.

Mishna. Chapters 1 and 3 of the fifth tractate Makkot of the fourth order Nezikin in the Mishna. Retrieved from http://halakhah.com/pdf/nezikin/Makkoth.pdf.

Mitchell, W. C. (1927). *Business Cycles, the Problem and Its Setting*. New York: National Bureau of Economic Research. Retrieved from http://www.nber.org/chapters/c0682.pdf.

Mnookin, S. (2011). *The Panic Virus: A True Story of Medicine, Science, and Fear*. New York: Simon & Schuster.

Montague, P. R., Berns, G. S., Cohen, J. D., McClure, S. M., Pagnoni, G., Dhamala, M., Wiest, M. C., et al. (2002). Hyperscanning: Simultaneous fMRI during linked social interactions. *NeuroImage*, *16*(4), 1159–64.

Morris, J. S., Öhman, A., & Dolan, R. J. (1999). A subcortical pathway to the right amygdala mediating "unseen" fear. *Proceedings of the National Academy of Sciences*, *96*(4), 1680–1685.

Myer, D. S. (n.d.). *Semiannual Report of the War Relocation Authority, for the Period January 1 to June 30, 1946* (p. 29). Washington, DC. Retrieved from http://www.trumanlibrary.org/whistlestop/study_collections/japanese_internment/documents/index.php?pagenumber=4anddocumentid=62anddocumentdate=1946-00-00andcollectionid=JIandnav=ok.

Nagamo, J., Arimoto, S., and Yoshizawa, S. (1962). An active pulse transmission line simulating nerve axon. *Proceedings of the IRE, 50*(10), 2061–2070. doi:10.1109/PGEC.1963.263454

Nielsen, J., and Shapiro, S. (2009). Coping with fear through suppression and avoidance of threatening information. *Journal of Experimental Psychology Applied, 15*(3), 258–274.

Norrholm, S. D., Jovanovic, T., Vervliet, B., Myers, K. M., Davis, M., Rothbaum, B. O., and Duncan, E. J. (2006). Conditioned fear extinction and reinstatement in a human fear-potentiated startle paradigm. *Learning & Memory*, *13*(6), 681–685.

Öhman, A. (2005). The role of the amygdala in human fear: Automatic detection of threat. *Psychoneuroendocrinology*, *30*(10), 953–8. doi:10.1016/j.psyneuen.2005.03.019

Öhman A., and Soares, J. J. (1994). "Unconscious anxiety": Phobic responses to masked stimuli. *Journal of Abnormal Psychology. 103*(2): 231–40.

Öhman, A., and Wiens, S. (2003). On the automaticity of autonomic responses in emotion: An evolutionary perspective. In R. J. Davidson, K. R. Scherer, and H. H. Goldsmith (Eds.), *Handbook of Affective Sciences*. New York: Oxford University.

O'Keefe, J., and Nadel, L. (1978). *The Hippocampus as a Cognitive Map*. Oxford: Clarendon Press.

Olsson, A., Nearing, K. I., and Phelps, E. A. (2007). Learning fears by observing others: The neural systems of social fear transmission. *Social, Cognitive, and Affective Neuroscience*, *2*(1), 3–11.

Olsson, A., and Öhman, A. (2009). The affective neuroscience of emotion: Automatic activation, interoception, and emotion regulation. In G. Berntson and J. T. Cacioppo (Eds.), *Handbook of Neuroscience for the Behavioral Sciences*. Vol. 2. (pp. 731–744). New York: Wiley.

Olsson, A., and Phelps, E. A. (2004). Learned fear of "Unseen" faces after Pavlovian, observational, and instructed fear. *Psychological Science 12*, 822–828.

Olsson, A., and Phelps, E. A. (2007). Social learning of fear. *Nature Neuroscience, 10*(9), 1095–1102.

Orwell, George (1949). *Nineteen Eighty-Four. A Novel.* London: Secker & Warburg.

Parker, J., and Epstein, J. M. (2011). A distributed platform for global-scale agent-based models of disease transmission. *ACM Transactions on Modeling and Computer Simulation, 22*(1), 1–25. doi:10.1145/2043635.2043637

Patterson, S. L., Abel, T., Deuel, T.A.S., Martin, K. C., Rose, J. C., and Kandel, E. R. (1996). Recombinant BDNF rescues deficits in basal synaptic transmission and hippocampal LTP in BDNF knockout mice. *Neuron, 16*(6), 1137–1145.

Paulos, J. A. (2000). Counting body parts. Review of *The Mathematical Brain*, by Brian Butterworth. *London Review of Books*, 27–28.

Pavlov, I. P. (1903). The experimental psychology and psychopathology of animals. *The 14th International Medical Congress*. Madrid.

Pavlov, I. P. (1927; 1960 Edition). *Conditioned Reflexes; An Investigation of the Physiological Activity of the Cerebral Cortex*. New York: Dover.

Pearce, J. M., and Hall, G. (1980). A model for Pavlovian learning: Variations in the effectiveness of conditioned but not of unconditioned stimuli. *Psychological Review, 87*(6), 532–552.

Pennington, N., and Hastie, R. (1991). A cognitive theory of juror decision making: The story model. *Cardozo Law Review, 13*(2–3), 519–556.

Penrod, S., and Hastie, R. (1980). A computer simulation of jury decision making. *Psychological Review, 87*(2), 133–159. doi:10.1037/0033-295X.87.2.133

Penrose, R. (1999). *The Emperor's New Mind: Concerning Computers, Minds, And The Laws Of Physics*. Oxford: Oxford University Press.

Phelps, E. A., Delgado, M. R., Nearing, K. I., and LeDoux, J. E. (2004). Extinction learning in humans: Role of the amygdala and vmPFC. *Neuron, 43*(6), 897–905.

Phelps, E. A., and LeDoux, J. E. (2005). Contributions of the amygdala to emotion processing: from animal models to human behavior. *Neuron, 48*(2), 175–87. doi:10.1016/j.neuron.2005.09.025

Philipson, T. (2002). Economic Epidemiology and Infectious Disease. In J. Newhouse and T. Culyer (Eds.). *Handbook of Health Economics*, (pp. 1761–1799). New York: North-Holland.

Piazza, M., Mechelli, A., Butterworth, B., and Price, C. J. (2002). Are subitizing and counting implemented as separate or functionally overlapping processes? *Neuroimage, 15*(2): 435–446.

Pittam, J., and Scherer, K. R. (1993). Vocal expression and communication of emotion. In M. Lewis and J. Haviland (Eds.), *The Handbook of Emotions* (pp. 185–198). New York: Guilford.

Pratkanis, A. R., and Aronson, E. (2001). *Age of Propaganda: The Everyday Use and Abuse of Persuasion*. New York: Freeman.

Price, D.D.S. (1976). A general theory of bibliometric and other cumulative advantage processes. *Journal of the American Society for Information Science, 27*(5), 292–306. doi:10.1002/asi.4630270505

Quinn, S. C., Thomas, T., and Kumar, S. (2008). The anthrax vaccine and research: Reactions from postal workers and public health professionals. *Biosecurity and Bioterrorism: Biodefense Strategy, Practice, and Science, 6*(4), 321–333.

Quinn, S. C., Thomas, T., and McAllister, C. (2005). Postal workers' perspectives on communication during the anthrax attack. *Biosecurity and Bioterrorism: Biodefense Srategy, Practice, and Science*, *3*(3), 207–215.

Railsback, S. F., and Grimm, V. (2011). *Agent-Based and Individual-Based Modeling: A Practical Introduction*. Princeton, NJ: Princeton University Press.

Rescorla, R. A. (1988). Pavlovian conditioning. It's not what you think it is. *American Psychologist* , *43*(3), 151–160.

Rescorla, R. A., and Heth, C. D. (1975). Reinstatement of fear to an extinguished conditioned stimulus. *Journal of Experimental Psychology: Animal Behavior Processes*, *1*(1), 88–96.

Rescorla, R. A., and Wagner, A. R. (1972). A theory of Pavlovian conditioning: Variations in the effectiveness of reinforcement and nonreinforcement. In A. H. Black and W. F. Prokasy (Eds.), *Classical Conditioning II: Current Research and Theory* (Vol. 20, pp. 64–99). New York: Appleton-Century-Crofts.

Resnick, M. 1994. *Turtles, Termites, and Traffic Jams: Explorations in Massively Parallel Microworlds*. Cambridge: MIT Press.

Rizzolatti, G., and Craighero, L. (2004). The mirror-neuron system. *Annual Review of Neuroscience*, *27*(1), 169–92. doi:10.1146/annurev.neuro.27.070203.144230

Rizzolatti, G., and Fabbri-Desto, M. (2009). The mirror neuron system. In G. G. Berntson and J. T. Cacioppo (Eds.), *Handbook of Neuroscience for the Behavioral Sciences* (pp. 337–357). Hoboken, NJ: Wiley.

Rodrigues, S. M., LeDoux, J. E., and Sapolsky, R. M. (2009). The influence of stress hormones on fear circuitry. *Annual Review of Neuroscience*, *32*(March), 289–313.

Rosenblatt, F. (1962). *Principles of Neurodynamics: Perceptrons and the Theory of Brain Mechanism*. Washington, DC: Spartan Books.

Rumelhart, D. E., and McClelland, J. L. (1987). *Parallel Distributed Processing: Explorations in the Microstructure of Cognition. Foundations*. Cambridge: MIT Press.

Schelling, T. C. (1971). Dynamic models of segregation. *Journal of Mathematical Sociology*, *1*(2), 143–186.

———. (1978). *Micromotives and Macrobehavior*. New York: WW Norton & Company.

Schiller, D., Cain, C. K., Curley, N. G., Schwartz, J. S., Stern, S. A., LeDoux, J. E., and Phelps, E. A. (2008). Evidence for recovery of fear following immediate extinction in rats and humans. *Learning & Memory*, *15*(6), 394–402.

Schiller, D., Levy, I., Niv, Y., LeDoux, J. E., and Phelps, E. A. (2008). From fear to safety and back: Reversal of fear in the human brain. *Journal of Neuroscience*, *28*(45), 11517–11525.

Schneider, W. (2003). Controlled and automatic processing: Behavior, theory, and biological mechanisms. *Cognitive Science*, *27*(3), 525–559. doi:10.1016/S0364-0213(03)00011-9

Seasonal price fluctuations for fresh fruit and vegetables. (n.d.). *Statistics New Zealand Tatauranga Aotearoa*. Retrieved December 21, 2012, from http://www.stats.govt.nz/browse_for_stats/economic_indicators/prices_indexes/seasonal-fluctuations-in-fresh-fruit-and-vegetables.aspx.

Sehlmeyer, C., Schöning, S., Zwitserlood, P., Pfleiderer, B., Kircher, T., Arolt, V., and Konrad, C. (2009). Human fear conditioning and extinction in neuroimaging: A systematic review. *PLoS ONE*, *4*(6), 16.

Shalizi, C. R., and Thomas, A. C. (2011). Homophily and contagion are generically confounded in observational social network studies. *Sociological Methods & Research*, *40*(2): 211–239.

Sigurdsson, T., Doyère, V., Cain, C. K., and LeDoux, J. E. (2007). Long-term potentiation in the amygdala: A cellular mechanism of fear learning and memory. *Neuropharmacology*, *52*(1), 215–227.

Simon, H. A. (1955). On a class of skew distribution functions. *Biometrika*, *42*(3–4), 425–440. doi:10.1093/biomet/42.3-4.425

———. (1982). *Models of Bounded Rationality*. Cambridge: MIT Press.

Slovic, P. (1995). The construction of preference. *American Psychologist*, *50*(5), 364–371. doi:10.1037/0003-066X.50.5.364

Slovic, P. (2007). If I look at the mass I will never act: Psychic numbing and genocide. *Judgment and Decision Making*, *2*(2), 79. doi:10.1007/978-90-481-8647-1

Slovic, P., Finucane, M., Peters, E., & MacGregor, D. G. (2002). The affect heuristic. In T. Gilovich, D. Griffin, & D. Kahneman (Eds.), *Heuristics and biases: The psychology of Intuitive Judgment* (pp. 397–420). New York: Cambridge University Press.

Slovic, P., and Lichtenstein, S. (1968). Relative importance of probabilities and payoffs in risk taking. *Journal of Experimental Psychology*, *78*(3), 1–18.

Slovic, P., Monahan, J., and MacGregor, D. G. (2000). Violence risk assessment and risk communication: The effects of using actual cases, providing instruction, and employing probability versus frequency formats. *Law and Human Behavior*, *24*, 271–296.

Small, D. A., Lerner, J. S., and Fischhoff, B. (2006). Emotion priming and attributions for terrorism: Americans' reactions in a national field experiment. *Political Psychology*, *27*(2), 289–298. doi:10.1111/j.1467-9221.2006.00007.x

Small, G. W., Feinberg, D. T., Steinberg, D., and Collins, M. T. (1994). A sudden outbreak of illness suggestive of mass hysteria in schoolchildren. *Archives of Family Medicine*, *3*(8), 711–716.

Smith, A. (1759; 1982 Edition). *The Theory of Moral Sentiments*. D. D. Raphael and A. L. Macfie (Eds.). The Glasgow Edition of the Works and Correspondence of Adam Smith, 1. Indianapolis: Liberty Fund. Retrieved from http://oll.libertyfund.org/title/192.

Smith, E. R., and DeCoster, J. (2000). Dual-process models in social and cognitive psychology: Conceptual integration and links to underlying memory systems. *Personality and Social Psychology Review*, *4*(2), 108–131. doi:10.1207/S15327957PSPR0402

Spinoza, B. de. (1677; 2007 Edition). *The Ethics*. The Project Gutenberg. EBook 380. Retrieved from http://www.gutenberg.org/files/3800/3800-h/3800-h.htm.

Spitzer, M., Fischbacher, U., Herrnberger, B., Grön, G., and Fehr, E. (2007). The neural signature of social norm compliance. *Neuron*, *56*(1), 185–196.

Stanovich, K. E., and West, R. F. (2000). Individual differences in reasoning: implications for the rationality debate? *Behavioral and Brain Sciences*, *23*(5), 645–665; discussion 665–726.

Strack, F., Martin, L. L., and Stepper, S. (1988). Inhibiting and facilitating conditions of the human smile: A nonobtrusive test of the facial feedback hypothesis. *Journal of Personality and Social Psychology*, *54*(5), 768–777.

Strogatz, S. H. (2001). *Nonlinear Dynamics and Chaos: With Applications to Physics, Biology, Chemistry and Engineering*. Cambridge, MA: Perseus Books.

Sunstein, C. R., & Hastie, R. (2008). Four failures of deliberating groups. John M. Olin Law and Economics Working Paper no. 401. University of Chicago.

Sutton, R. S., and Barto, A. G. (1981). Toward a modern theory of adaptive networks: Expectation and prediction. *Psychological Review*, *88*(2), 135–170.

———. (1987). A temporal-difference model of classical conditioning. *Proceedings of the Ninth Annual Conference of the Cognitive Science Society, GTE TR87 5*, 355–378.

———. (1998). *Reinforcement Learning: An Introduction.* Bradford Books. Cambridge: MIT Press.

Swahn, M. H., Mahendra, R. R., Paulozzi, L. J., Winston, R. L., Shelley, G. A., Taliano, J., Frazier, L., et al. (2003). Violent attacks on Middle Easterners in the United States during the month following the September 11, 2001, terrorist attacks. *Injury Prevention Journal of the International Society for Child and Adolescent Injury Prevention, 9*(2), 187–189.

Tamietto, M., Castelli, L., Vighetti, S., Perozzo, P., Geminiani, G., Weiskrantz, L., and De Gelder, B. (2009). Unseen facial and bodily expressions trigger fast emotional reactions. *Proceedings of the National Academy of Sciences, 106*(42), 17661–17666.

Telzer, E. H., Humphreys, K. L., Shapiro, M., and Tottenham, N. (2012). Amygdala sensitivity to race is not present in childhood but emerges over adolescence. *Journal of Cognitive Neuroscience.* doi:10.1162/jocn_a_00311

Tesfatsion, L., and Judd, K. L. (Eds.). (2006). *Handbook of Computational Economics, Volume II: Agent-Based Computational Economics.* Amsterdam: Elsevier.

Tolstoy, L. (1869; 1998 Edition). *War and Peace.* A. Mandelker, A. Maude, and L. S. Maude (Eds.). Oxford: Oxford University Press.

Tooby, J., and Cosmides, L. (2008). The evolutionary psychology of the emotions and their relationship to internal regulatory variables. *Evolutionary Psychology,* 114–137.

Tucker, R. C., ed. (1972). *The Marx-Engels Reader.* New York: Norton.

Tversky, A., and Kahneman, D. (1971). Belief in the law of small numbers. *Psychological Bulletin, 76*(2), 105–110. doi:10.1037/h0031322

———. (1974). Judgment under uncertainty: Heuristics and biases. *Science, 185*(4157), 1124–31. doi:10.1126/science.185.4157.1124

———. (1982). Evidential impact of base rates. In D. Kahneman, P. Slovic, and A. Tversky (Eds.), *Judgment under Uncertainty: Heuristics and Biases* (pp. 153–160). Cambridge: Cambridge University Press.

Using Social Norms to Attack Prostate Cancer among African Americans. (n.d.). Pittsburgh. Retrieved from http://ncmhd.nih.gov/spotlight/prostate.asp.

Vul, E., Harris, C., Winkielman, P., and Pashler, H. (2009a). Puzzlingly high correlations in fMRI studies of emotion, personality, and social cognition. *Perspectives on Psychological Science, 4*(3), 274–290. doi:10.1111/j.1745-6924.2009.01125.x

———. (2009b). Reply to comments on "Puzzlingly High Correlations in fMRI Studies of Emotion, Personality, and Social Cognition." *Perspectives on Psychological Science, 4*(3), 319–324. doi:10.1111/j.1745-6924.2009.01132.x

Wagenaar, D. J. (1996). Fourier transform introduction. *Joint Program in Nuclear Medicine: PHYSICS.*

Wang, G. J., Volkow, N. D., Logan, J., Pappas, N. R., Wong, C. T., Zhu, W., Netusil, N., et al. (2001). Brain dopamine and obesity. *Lance, 357*(9253), 354–357.

Watson, J. B., and Rayner, R. (1920; 2000 Reprint). Conditioned emotional reactions. *American Psychologist, 55*(3), 313–317. Reprinted article. Originally appeared in *Journal of Experimental Psychology*, 1920, *3*, 1–14.

Watts, D. J., and Strogatz, S. H. (1998). Collective dynamics of "small-world" networks. *Nature, 393*(6684), 440–2. doi:10.1038/30918

Whalen, P. J., Rauch, S. L., Etcoff, N. L., McInerney, S. C., Lee, M. B., and Jenike, M. A. (1998). Masked presentations of emotional facial expressions modulate amygdala activity without explicit knowledge. *Journal of Neuroscience, 18*(1), 411–418.

Wiggins, S. (1990). *Introduction to Applied Nonlinear Dynamical Systems and Chaos.* New York: Springer-Verlag.

Young, H. P. (1993). The evolution of conventions. *Econometrica, 61*(1), 57–84. doi:10.2307/2951778

———. (1998). *Individual Strategy and Social Structure.* Princeton, NJ: Princeton University Press.

Yule, G. U. (1925). A mathematical theory of evolution, based on the conclusions of Dr. J. C. Willis, F.R.S. *Philosophical Transactions of the Royal Society of London. Series B, Containing Papers of a Biological Character, 213*(402–410), 21–87. doi:10.1098/rstb.1925.0002

Zillmann, D., Bryant, J., Cantor, J. R., and Day, K. D. (1975). Irrelevance of mitigating circumstances in retaliatory behavior at high levels of excitation. *Journal of Research in Personality, 9*(4), 282–293. doi:10.1016/0092-6566(75)90003-3

Index

18th Brumaire of Agent_Zero (extension), 165–68, 192

action: binary, *xiii,* 6–7; imitation and, 2, 8, 59–60, 89; initiation of, 11–12, 55–56, 71–72, 80, 94–96, 187–88; rule, 6–7, 48n84; threshold, 7, 36–37, 46–48, 53–54
Adolphs, R. (et al.), 26, 105
affect heuristic, 123
affective component of *Agent_Zero,* 1–2, 5, 84–85; coupling with cognitive component, 122–27. *See also* fear; Rescorla-Wagner model
affinity trajectories, 133–35
age and impulse control (extension), 109–10
agent-based models, 81; described, 11–12; heterogeneity and, 63; individual-based models, 72; multiple agent layers, 160–65; neural deepening and, 185; reproduction, 165. *See also* three-agent model; two-agent model
Agent_Zero, xii; components of, 3, 9, 14 (*See also* affective component, cognitive component, social component); coupling between components, 122–27; dispositional contagion and, *xiii,* 2–3, 59–63; dispositional trajectory of, 55n93; future applications of, 17–18, 181–92; as initiator, 94–96; instances of general class, 9; movies, 16; *NetLogo* Applets, 16; *NetLogo* source code, 17; Rescorla-Wagner model and, *xiii,* 33 (*See also* Rescorla-Wagner model); Slaughter of the Innocents, 90–94; spatial sampling radius, 96–97; vision, 96–99
Albert, R., 159
amygdala, 4, 103n146; fear conditioning and, 19, 21–26, 31–32, 61–62; inputs to, 21–23, 31, 62n102; lesion studies and, 46, 102–6; outputs, 24–25; prejudice and, 38; and recognition of fearful facial expressions, 26, 105
anchoring and adjustment, 46, 121–22, 182–83
animal models, 42–44, 85
Arab Spring extension, 14–15, 138–42; homophily and, 128, 160, 188; leaderless revolutions, 142; social media and, 62–63, 138–42, 188
areal bias, 86
Aristotle, 1n4
artificial societies. *See* agent-based models
Arzi, A (et. al), 26
Asch, S. E., 63–65
associative strength, 30–31, 34
attack rate: code block for, 177n212
auditory stimuli, 11, 21–23, 62n102
automatic and controlled cognition, 5n11. *See also* "low and high road"
Aviezer, H., 59
Axelrod, Robert, 38–39, 81, 118n81
Axtell, Robert, *xi,* 120n164

backward masking, 26, 37
Barábasi, Albert-László, 159
Barto, A. G., 44
base rate neglect, 86
Bauer, E. P., 27–28
Bayesian updating, 50, 182
Berns, G. S. (et al.), 65
betrayal, 37, 40, 79
binge eating, 13, 80, 108, 189–90
blocking and selective discrimination, 38–40
Bloom, F., 12n32, 24–25, 43
BOLD signal, 32
bounded rationality, 9, 46, 50–51, 88
brain regions, 4–5
Bucy, P. C., 102
Butterworth, B., 50, 86
bystander effects, 14, 116–18

Cacioppo, John, 26, 58, 60, 63, 98
capital flight, 188–89
Centola, D., 73
civil violence, 37n63, 176, 187–88, 192
Code Library, 16–17
cognitive component of *Agent_Zero,* 1–2, 85–88; bounded rationality and, 9, 46–52, 88; coupling with affective component, 122–27; fear conditioning and self-control, 29, 40–41, 188; memory and, 49–50; sample selection error and, 49; social influence and, 188
cognitive realism, *xi–xii*
combat, 99–100

communication: postural cues and, 59–60; social media, 11, 14–15, 62–63, 89, 128, 138, 139–42, 188; verbal, 60
complex contagion, 73
computational social neuroscience, 98
Conditioned Reflexes (Pavlov), 28–29
conditioning: classical, 19–20; defined, 19–20; Little Albert, 34; nomenclature of, 29–30; observational, 60; trials, 31–32. *See also* fear conditioning; learning
conformity, social, 11, 16; Asch experiment, 63–65; conformist empirical estimates, 63–67; and democracy, 67; enforcement of, 65; evolutionary advantage, 66–67; jury processes and, 127; neural basis for, 65–67; pain and nonconformity, 65–66, 190
conformity effects: neural bases for, 65–66, 190
connection strength: affective homophily and, 128–30; affinity trajectory and, 133–34; in classical network theory, 15; general setup for, 130–35; and heterogeneous exponents, 132–33; homophily and, 128–29; interagent weight and, 15; learning rates and, 134–35; Rescorla-Wagner setup, 131–32; weight surface and, 129–30
contagion: complex, 73, 96; Kermack-McKendrick disease-transmission model, 33. *See also* dispositional contagion
coordination game, 44–45
Cosmides, L., 5n11
coupling: between *Agent_Zero* components, 122–27; coupled trajectories, 9–10; homophily and coupling strength, 14
Cunningham, W. A. (et al.), 38

Darwin, Charles, 25, 51
defensive aggression, 12n32
de Gelder, B. (et al.), 59
deliberation. *See* cognitive component of *Agent_Zero*
Desteno, D., 123
destructive radii: endogenous, 107–9
disposition, *xiii,* 3; action rule and, 6–7; interdependent trajectories, 73; in *NetLogo* graphical output, 88; skeletal equation, 7–9. *See also* net disposition; solo disposition; total disposition
dispositional contagion, *xiii,* 2–3; *vs.* behavioral imitation, 2–3, 7–9; complex contagion and, 73; escalation and, 91; facial and postural mimicry and, 26, 59–60; heterogeneity of, 63; leadership as susceptibility to, 94–96, 187–88; mechanisms of, in *Agent_Zero,* 59–63; range contrasted with spatial sampling radius, 88–89, 94; in The Slaughter of Innocents, 90–94; in two-agent model, 198–202

eating behavior, 13, 80, 108, 189–90
Eaton, William, 59
economics, 13; capital flight, 16, 114, 188–89; financial panics, 58–59; marketing, 16, 173–76, 189; prices, introduction of, 13, 16, 168–72, 175–76, 189; seasonal demand and, 172–74, 189. See also *Homo economicus*
edges, 15, 129, 152
Einstein, Albert, 5
elections, 152n 182
emergency first responders, 98–99
emergentism, 71–72, 152–59, 184
emotional amplification of probability estimates, 122–23
Emotional Contagion (Hatfield et al.), 58
emotions. *See* affective component of *Agent_Zero*
empathy, 185
An Enquiry Concerning Human Understanding (Hume), 30–31
entanglement of passion and reason, 13–14, 122–27
Epstein, J., 58, 192n223
Epstein, S. D., *xi,* 4n10, 53n92
escalation, 192; dispositional contagion and, 91; spirals, 13, 15, 176–80, 185
evolution, biological: conformity and, 66–67; of fear circuitry, 20–22, 25–26, 28–29, 43; reflexive responses and, 28–29, 31; social acquisition of fear as evolutionary advantage, 28, 57, 61–62
evolution, of network structures, 14–15, 159; and political attitudes, 159–60
The Expression of Emotions in Man and Animals (Darwin), 25
extensions of *Agent_Zero* model: listed, 13
extinction: affinity dynamics and, 136; fear conditioning and, 27–29, 45; half-life of, 44–45; in jury process extension, 146; of majorities, 78; *Mathematica* 8.0 code for extinction phase, 206–7; and mPFC, 45; posttraumatic stress and persistence of disposition, 45–46, 78–79; in Rescorla-Wagner model, 44–45, 56n97, 84n122, 104n148

Extraordinary Popular Delusions and the Madness of Crowds (Mackay), 57–58

fear: amygdala and, 19, 21–26, 31–32, 61–62; circuitry, 20–29, 43–44, 61–62; conscious deliberation and, 29, 40–41, 46–48, 188; contagion of, 26, 28–29, 59–60, 188; delayed, 26–27; direct stimulus and, 60; extinction of, 27–29, 45; maladaptive, 28–29; mirror neurons and observational acquisition of, 28, 62; mutual amplification dynamics, 126; novel threats and adaptive plasticity, 27; prejudices and, 28; retention of, 27–28. *See also* fear conditioning
fear conditioning, *xiii,* 34; amygdala and, 19, 21–26, 31–32, 61–62; extinction, 27–29, 45; "instructed," 10; observational, 60–63; and Pavlovian associative learning, 28–29; propaganda and, 28–29; and Rescorla-Wagner model, 33–37; unconscious, 19, 25–29, 40–41, 59, 188
fight *vs.* flight, 110–14; Case 1: Fight, 111; Case 2: Flight, 112–13; defensive aggression, 12n32; financial flight, 113
fitness, perils of, 20–21, 28–29
Fitzhugh-Nagamo model, 20
fMRI, interpretation of, 32
future research and applications, 17–18, 165, 179–92

game theory, 3, 44–45, 120n164, 174n205
Gazzola, V. (et al.), 62
generality of *Agent_Zero* model, 12
generative minimalism, 4, 5
Generative Social Science: Studies in Agent-Based Computational Modeling (J.M. Epstein), *xi*
Global Scale Agent Model (GSAM), 184–85
Growing Artificial Societies: Social Science from the Bottom Up (J.M. Epstein and R. Axtell), *xi*

half-life, 44–45
Hammond, R. A., 38–39
happiness, 100–102
Hart, A. J. (et al.), 39
Hatfield, E., 58, 60, 63
health behavior: binge eating, 13, 80, 108, 189–90; vaccination, refusal of, 12, 38, 44, 80, 108, 189
Heaviside unit step functions, 7, 41, 152–53, 197
Hebb, Donald, 19–20, 33, 40n70
Hebbian plasticity, 20, 33, 40
heuristics, *xiii,* 12, 46–50, 59, 86, 121, 123
hippocampus, 4, 23, 85, 118
Hobbes, Thomas, 47
Hodgkin-Huxley model, 20
Hofstadter, Richard, 40
Homo economicus, 91, 192–93
homophily, affective, 128–30; and Arab Spring extension, 128, 160, 188; in jury processes, 147–50; and network emergence, 16, 152, 155, 159–60; and network structure, 16, 134–35, 152–53, 158–60, 187, 191; social media and, 14–15, 140, 188
hormones, 21–22
"hot" and "cold" cognition, 5n11
Hume, David, 1, 46, 47; and "association of ideas," 30–31
hysteresis, 134–35

idealization, theory and, 192–93
imitation, 5, 8; action and, 2, 8, 59–60, 89; dispositional contagion as distinct from, 2–3, 7–9; influence as distinct from, 88–89; mirror neurons and, 28, 62; succession as distinct from, 70–71
impulse control: aging and, 109–10, 190; fear conditioning and, 29; Raskolnikov and, 186
individual-based models, 72
information: costs of, 98
inputs: amygdala, 21–23, 27, 103; "low and high" road, 23–24, 26, 31, 51
intergenerational transmission, 165–68
internment, 38–40, 52

James, William, 26–27, 59
Jamesian fear contagion mechanism, 59–60
Jasmine resolutions, 14–15, 138–42
Judd, K. L., 81
jury processes extension, 15–16, 127; change of venue, 149–52; courtroom trial phase, 145–47; extinction and, 146; impartiality, 146, 149; memory apparatus and, 143, 146–47; pretrial public phase, 143–45; sequestration phase, 147–52
justification *vs.* explanation, 18

Kahneman, D., 5n11, 49, 59
Kaldor, Nicholas, 185
Kamin, L. J., 40
Kandel, Eric, 118

Kanizsa, Gaetano, 86–87
Kermack-McKendrick disease-transmission model, 33
Kindleberger, Charles, 58–59
Klüver, H., 102
Knierim, J. J., 85
"knock-out" studies, 102–4, 191
Kross, E. (et al.), 65–66

Latané-Darley experiment, 13–14, 115–16, 190; *Agent_Zero* replication of, 185; flight and, 113, 114; three-agent model and replication of, 114–15, 185; threshold imputation and, 8n23, 13–14, 114–18, 185–90, 195–96; Zillman and, 190
Lazar, D., 159–60
Lazerson, A., 12n32, 24–25, 43
leadership: Agent 0 and, 94–96; Arab Spring as leaderless, 142; disposition of initiating agent, 11, 55n93, 80; distinct from initiation, 12, 55–56, 63, 71–72, 80, 187–88; "Great Man Theory," 71; as susceptibility to dispositional contagion, 94–96, 187–88; swarmocracy and, 71–72; in two-agent model, 55–56
learning: defined, 34; Hebbian plasticity and, 20, 33, 40; intergenerational transmission of memory/narrative, 165–68, 192; and network connection strength, 134–35; Pavlovian associative, 5, 34–35; rates of, 34–35, 134–35; in Rescorla-Wagner model, 5, 69; and salient surprise in Rescorla-Wagner model, 34–40
Le Bon, Gustave, 57
LeDoux, J. E., 43–45, 50–51, 57, 59–61; on auditory stimulus and the amygdala, 21; on fear circuitry, 19–21; "low road and high road," 23–24; on physiological responses to fear, 22; unconscious activation of fear, 25–28
Lerner, J. S., 123
lesioning *Agent-Zero,* 103–5, 191
lesion studies, 19n1, 105; code block to "knock-out" *Agent_Zero* amygdala, 103–5, 191; "knock-out" animal models, 102–3
Levitan, L. C., 160
Lotka-Volterra equations, 44
"low and high road," 23–24, 26, 31, 51
lynching, 2–3

Mackay, Charles, 57–58
Macy, M., 73
majority rule, 68
Manias, Panics, and Crashes (Kindleberger), 58–59
marketing, 16, 189; neuromarketing, 175–76
Martin, L. L., 59
Marx, Karl, 165–68
Mathematica 8.0 code, 16–17, 197–211; extinction phase, 206–7; group's disposition trajectory (vector field excursion), 208; Rescorla-Wagner model, basics, 197; strength-homophily dynamics, 209–11; Three-Agent Runs, 202–7; Two-Agent Dispositional Contagion, 198–202
The Mathematical Brain (Butterworth), 50, 86
mathematical model: described, 10–11
McCarthyism or Joseph McCarthy, 40n69, 57–58
McNaughton, B. L., 85
memory: declarative, 118; episodic, 118; extension of *Agent_Zero* model, 118–22; hippocampus and, 118; jury processes and, 143, 146–47; moving average and, 121–22; moving median and, 122
Milgram experiment, 115
Miller, J. H., 81
mimicry. *See* imitation
mirror neurons, 28, 62
Mishna, 74–75
Mitchell, Wesley, 169, 172
mitigating circumstances: Zillman's Experiment, 93–94
modularity, 33n58
Moore neighborhoods, 48
mutual escalation, 13, 15, 176–80, 192; as "stylized dynamic," 185

Napoleon Bonaparte, 96, 142
Nearing, K. I., 60–62
negative radial moralities, 74–75
Nelson, C. A., 12n32, 24–25, 43
net disposition, 7–8, 54–56, 80, 94n137; and fight *vs.* flight, 113–14; in jury processes, 144, 147; and market economics, 174–75; *Mathematica* Code and, 206, 209; in *NetLogo* output, 88; threshold imputation and, 95n138, 115–16, 124n166
NetLogo Code, 16–17, 213–20; ticks as time, 90n131
network dynamics, 15
network formation, 128, 159, 191; internode strength and thresholds for, 152–53; preferential attachment and, 159, 185–86

network structure: and affinity dynamics, 158–59; defined and described, 158; emergent dynamics of, 152–59; homophily and formation of, 16, 152, 155, 159–60; multiple network membership, 160; as Poincaré map, 153–59. *See also* connection strength
network theory, 15
neural networks: backpropagation, 183; perceptron, 183
neuroeconomics, 3n7
neuromarketing, 175–76
neuroscience: and *Agent_Zero, xii,* 3–5, 184–85, 187, 193; of fear, 19–29, 60–63 (*See also* amygdala); fMRIs, 3–4, 19, 32, 65; lesion studies, 19n1, 102–5; of memory, 50, 118; social conformity and, 65–67, 189–90
Newell, Allen, 20n42
nodes, 15, 159
nonequilibrium dynamics, 135–38
norms, social: neural basis for conformity effects, 65–67; as vector fields, 74–78
numerical cartography, 181

obesity, 13
Occupy Wall Street, 15n35, 142
Öhman, A., 25, 28, 38
Olsson, A., 32, 38, 60–62
Organ of Corti, 119–20
outgroup prejudice and perceived threats, 37–38

Page, S. E., 81
pain of independence, 65–66, 190
Parable 1. *See* Slaughter of Innocents
Parable 2 (Agent_Zero initiates). *See* leadership, as susceptibility to dispositional contagion
parameter settings for model runs, 221–25
paranoia, 40
"The Paranoid Style in American Politics" (Hofstadter), 40
Parker, Jon, 184, 195
passions. *See* affective component of *Agent_Zero*
pattern perception, 86–87; generative aspect of, 87
Pavlov, Ivan, 29
Pavlovian associative learning, 5; fear conditioning and, 28–29; terminology of, 29–30
Pearce-Hall model, 44, 182–83
Pearl Harbor, 38–40
perceptron, 183
Petty, R. E., 98, 123
Phaedrus (Plato), 47
phantom edge phenomenon, 86–87
Phelps, Elizabeth A., 32, 34n60, 60–62
place cells, 85
plasticity, Hebbian, 19–20, 27, 33, 40n70, 41
Plato, allegory of the Charioteer, 47
Pleistocene man, 27, 40
politics: elections, 152n 182; paranoia and, 40; polarized, 124–25; social conformity as deterrent to dissent, 67
postural cues, 59–60
preferential attachment, 159, 185–86
prefrontal cortex (PFC), 23, 45–46
prejudice: fear conditioning and, 28; outgroup prejudice and perceived threats, 37–38
Price, D. De Solla, 159
prices, 13, 16, 168–72, 175–76, 189
Prisoners' Dilemma, 44–45, 174n205
probability estimation, *xiii,* 9n26, 12, 50; surprise and error in, 34, 37
propaganda: fear conditioning and, 28–30, 87
proximity-dependent weights, 183–84
pseudocode, 89
psychic numbing, 163n194, 185
PTSD (Posttraumatic stress disorder), 45–46, 78–79, 99–100, 190, 206

random movement (no activation), 221–25
Rapson, R. L., 58, 60, 63
Raskolnikov, *xi–xii,* 11–12, 186
rationality: bounded, 46, 50–51, 88; canonical, 46, 126; departure from, 46
reason: Hume and, 1. *See also* cognitive component of *Agent_Zero*
recursion, 165
replicability, 16–17, 81–82
representativeness heuristics, 49
Rescorla-Wagner model, *xiii,* 33–37; and affective component of *Agent_Zero,* 5, 9–10, 20, 44, 46, 84–85; alternatives to, 33n58, 182–83; classic setup (homogeneous learners), 69, 130–32; conditioning in, 5, 29–31, 33, 69; connection strength in, 131–32; extinction in, 44–45, 56n97, 84n122, 104n148; generalizing, 67–69; heterogeneous variant, 69–70, 132–35, 202–5; *Mathematica* 8.0 code for, 197; salient surprise in, 34–40
Resnick, M., 81
retaliation, 46, 93, 127, 176–80, 187–88

retention, 20, 27–28; inherited memory, 192. *See also* PTSD (Posttraumatic stress disorder)
revolutions, Jasmine, 14–15, 138–42
Rosenblatt, F., 183
Rucker, D. D., 123

Salem witch mania, 57–58
sample selection error, 49
scaling up *Agent_Zero,* 184–85
Schafe, G. E., 27–28
secondary somatosensory cortex, 65
Sehlmeyer, A. (et al.), 44
Shannon entropy, 124–25
Simon, Herbert, 46
skeletal equation, 7–8, 9–10, 11n28, 118, 124, 138; threshold imputation and, 115–16, 195
Slaughter of Innocents: civil violence and, 187–88; computational parable, 90–94; escalation in, 91; *NetLogo* Source Code, 213–20; Zillman's Experiment and, 93–94
Slovic, Paul, 123, 163n194, 185
Small, D. A., 123
Small, G. W. (et al), 59
small-world networks, 73, 183
Smith, Adam, 59
smoke-filled room. *See* Latané-Darley experiment
snakes, 5n11, 20–22, 45n79, 51
SOAR model, 20n42
social animals, 1; plurality of, 160. *See also* norms, social
social component of *Agent_Zero,* 1–2; conformity effects and, 63–67; coupled dispositional trajectories and, 52–53; dispositional contagion and, 1, 5, 55–60, 62–63, 70–71, 73; three agent model and, 68–71; Tolstoy's "swarm-life" and, 71–73
social dynamics: generation of, *xiii,* 2–4; in jury processes, 15, 147–52; recursion and, 165; scaling and, 184–85
social influence, 1, 13, 29, 80, 183. *See also* homophily, affective; norms, social
social media, 11, 14–15, 62–63, 128, 138–42
social networks, 9–10, 80, 128. *See also* homophily, affective
social psychology, 17, 185, 193. *See also* Latané-Darley experiment
The Sociology of Mental Disorders (Eaton), 59
solo disposition, 6, 10, 12; and three-agent model, 68–71
spatial sampling radii, 11, 48, 50, 85–89; *vs.* distance, 88–89, 94n135; proximity-dependent weights, 183–84; and "vision," 96, 110
Spinoza, Baruch de, 1, 160
Stanovich, K. E., 5n11
Stepper, S., 59
stereotyping, 183–84
stimuli. *See* inputs
stimulus overgeneralization, 38
Strack, F., 59
Strogatz, Steven H., 154, 183
Sugarscape model, *xi*
surprise (subject's prediction error): associative learning and, 34–35, 37; betrayal as, 40
survival circuits, 43
Sutton, R. S., 44
swarm life, 71–72, 95–96, 142
synapses, 19–22, 27–28, 33, 41–43
Syria, 15, 142, 176
System 1 and System 2, 5n11

T effect, 86–87
Tesfatsion, L., 81
The Theory of Moral Sentiments (Smith), 59
thought experiments, agent-based, 105
threats: amygdala and response to, 21–27; auditory stimuli and perceived, 21–22; outgroup prejudice and perceived, 37–38; T-effect and threat inflation, 86–87; "threat inflation," 80
three-agent model, 68, 115, 184–85, 202–7
threshold: action, 7, 36–37, 46–48, 53–54; link, 152
threshold imputation, 8n23, 14, 95n138, 115–18, 185, 190; bounds for, 195–96; bystander effects and, 14, 116–18
time: ticks in *NetLogo,* 90n131; time scales, 98–102
tipping point, 35–37
Todorov, A., 59
Tolstoy, Leo, 71–72, 77, 96
Tooby, J., 5n11
total disposition, 3, 6–8, 12, 80, 95, 138
Treatise of Human Nature (Hume), 1
Trope, Y., 59
Tuskegee syphilis study, 40, 44, 79
Tversky, A, 49
two-agent model: coupled model, 51–54; dispositional contagion in, 198–202; extinction in, 56–57

unconscious processing and response, 6n14, 119–20; delay between unconscious and conscious responses, 21–22; facial and postural mimicry, 59–60; fear conditioning and, 19, 25–29, 40–41, 188
unit step functions, 7, 41, 152–53, 197
Urbach-Wiethe disease, 105

vaccination, refusal of, 12, 38, 44, 53, 58, 80, 108, 189
Van der Worp, H. (et al.), 62
vector fields, 74–78, 208
vectors, 7–8
vision: heterogeneous, 97–98; spatial sampling radius, 96–97
von Neumann neighborhoods, 11, 48, 91, 96, 166

Wagenaar, D. J., 119–20
War and Peace (Tolstoy), 71–72, 96
Watson, James, 34
Watts, Duncan J., 183
website for *Agent_Zero* code and supplemental resources, 16–17
Wegener, D. T., 123
weights, 6–7; homophily and, 14, 122, 128–29, 135–38, 158–59, 191, 209; interagent, 14–15, 114, 133–35, 151; proximity-dependent, 183; trajectories of, 131–32, 135; weight surface, 129–30. *See also* connection strength
West, R. F., 5n11
Wiens, S., 25
Wisser, P. S., 160
Witness to History, agent as, 161–65
World Trade Center attacks, 37–38

xenophobia, 38

Yeats, William Butler, 16
Young, H. Peyton, 120n164

Zillman's Experiment, 93–94, 127, 185, 190